METHODS OF
DETERMINING E_h AND pH
IN SEDIMENTARY ROCKS

by

Gennadii Anatol'evich Solomin

Foreword by
N. G. Fesenko

Authorized Translation from the Russian by
Paul Porter Sutton
Professor of Chemistry
North Carolina State College

Springer Science+Business Media, LLC
1965

First Printing — March 1965

Second Printing — August 1967

The original Russian text was published by Nauka in Moscow in 1964

METHODS OF DETERMINING E_h AND pH IN SEDIMENTARY ROCKS

K METODIKE OPREDELENIYA
OKISLITEL'NO-VOSSTANOVITEL'NOGO POTENTSIALA
I pH OSADOCHNYKH POROD

К методике определения окислительно-восстановительного
потенциала и рН осадочных пород

Геннадий Анатольевич Соломин

Library of Congress Catalog Card Number 65-11958

ISBN 978-1-4899-4955-4   ISBN 978-1-4899-4953-0 (eBook)
DOI 10.1007/978-1-4899-4953-0

© 1965 Springer Science+Business Media New York

CONTENTS

Foreword.   5

Preface.   7

Chapter 1. Redox Potentials, pH, and Their Significance for Geochemistry   9

Chapter 2. Methods of Characterizing the Oxidation States of Sedimentary Rocks.   15

Chapter 3. Distinguishing Features of Redox Potential Measurements in Sedimentary Rocks.
Apparatus for Measurements in an Atmosphere of Nitrogen   19

Chapter 4. The Effect of Moisture   25

Chapter 5. The Time Required for Measuring Redox Potentials of Sedimentary Rocks.   29

Chapter 6. The Significance of the Preliminary Treatment of the Platinum Electrodes.   33

Chapter 7. The Accuracy and Reproducibility of Redox Potential Measurements
on Sedimentary Rocks.   39

Chapter 8. Methods for pH Determinations in Sedimentary Rocks   43

Conclusions.   47

Literature   49

FOREWORD

The redox potential determines the mobility of an element of variable valency and fixes its dispersion or localization in the earth's surface. It follows that this potential is a significant guide in prospecting and an equally important factor in determining origin of a deposit. Numerous attempts at measuring redox potentials in sedimentary rocks have proven unsuccessful and many have concluded that it is impossible to determine these potentials.

The author of the present work does not share this opinion and has developed a technique for such measurements which is fundamentally different from any proposed earlier. He has directed attention to the necessity of protecting the sample from the atmosphere and suggested various systems for measuring redox potentials under nitrogen. His studies on the role of moisture and the cause for delay in establishing the measured electrode potential are worthy of note. He has advanced an original solution of the problem of measurement in cases where more than 10-15 h are required for establishing an equilibrium redox potential value.

Particular attention has been given (Chapter 6) to the preliminary treatment of the platinum electrode. Here the author has brought clarity into confusion by showing the principal purpose of such treatment to lie in removal of adsorbed oxygen and reduction of the redox capacitance of the electrode. He points out that the oxygen saturation is adversely increased by the widely followed practice of treating the platinum electrode with chromate solutions, and proposes that this procedure be replaced by alkaline and cathodic polarization near the redox potential of the medium. His experiments show the superiority of low-inertia platinized glass and porcelain electrodes. It is clear that the lifetime of such electrodes could be considerably extended by resorting to plant-scale cathodic sputtering for depositing the platinum layer on the supporting solid.

The author points out deficiencies in the proposed technique of measuring redox potentials and suggests further improvements of the method.

He also suggests a more refined method for pH measurements on rocks, pointing to the increase in pH resulting from the presence of moisture and attempting to develop an explanation for it.

This is, on the whole, a valuable contribution to the study of the oxidation states of sedimentary rocks and the results reported here will undoubtedly find wide application in practice.

N. G. Fesenko

PREFACE

A dissertation on the oxidation states of water and rocks at the site of the Volgograd hydroelectric station which was written by the author under the direction of P. A. Kryukov, Candidate in Chemical Sciences, in 1960 has served as the basis for the present work. Two facts account for the publication of this material as a separate brochure. First, redox potentials (E_h) of sedimentary rocks continue to be measured by outmoded methods and are incorrect in many cases. Second, statements appearing in the literature suggest that the measurement of redox potentials in rocks is impossible in principle. The fact that statements of this kind have been left unanswered has doubtless impeded the development of methods for determining E_h, the principal characteristics of the oxidation state of a sedimentary rock.

The present work does not give a complete presentation of the technique of pH and E_h measurements, but treats only such inadequately understood methodological problems as the value of isolating the rock from the air, and the effect of moisture, the rate of establishing the electrode potential, and the preliminary electrode treatment on measurements in sedimentary rocks. The volume of the work has been kept to a minimum by bypassing certain theoretical questions and eliminating discussion of measuring devices, a topic which is well treated in textbooks on physical chemistry and electrochemistry, and in handbooks on physical measurements.

The methods proposed here for determination of E_h and pH were developed and successfully tested on Paleogene (Buchak and Tsaritsyn) rocks of various degrees of consolidation from the site of the Volgograd hydroelectric station and on fine grained clays from the site of the Saratov hydroelectric station. The natural moisture content of these rocks was 15-30% and their oxidation states were determined by the minerals pyrite, siderite, and iron silicate. Testing is needed to confirm the applicability of these methods to highly consolidated rocks of lower moisture content.

CHAPTER 1

REDOX POTENTIALS, pH, AND THEIR SIGNIFICANCE
FOR GEOCHEMISTRY

The state of a polyvalent element in a sedimentary rock or subsurface water is inseparately bound up with its redox potential and the acidity and alkalinity of the environment. These factors fix the minerological composition of the rock and the tendency of its various elements to migration. The study of redox potentials and acid–alkali equilibria is, therefore, one of the principal problems of geochemistry.

The redox potential gives a significant characterization of the oxidation state, while the pH characterizes the acid–alkali equilibrium.

The redox potential is the potential arising at a noble metal electrode immersed in a solution containing the redox system and inert with respect to it.* There are no means available for determining absolute potentials and electrode potentials are therefore usually measured against the normal hydrogen electrode, which is arbitrarily assigned zero potential. Relative redox potentials measured in this way will be designated by E_h.

The value of E_h is related to the activities of the various ionic forms of the polyvalent element through the well-known equation

$$E_h = E^\circ + \frac{RT}{nF} \ln \frac{a_{Ox}}{a_{Red}},$$ (1)

E° being the normal redox potential of the system, i.e., the potential measured at $a_{Ox} = a_{Red} = 1$, and a_{Ox}, and a_{Red} the respective activities of the oxidized and reduced forms of the element responsible for the reversible potential.

The most extensive table of normal potentials is that given by W. M. Latimer [1].

Equation (1) is the form of redox potential relation generally used in Europe. A different sign convention, namely:

$$E_{hAM} = -E_h \quad \text{and} \quad E^\circ_{AM} = -E^\circ,$$

has been widely accepted in America; this leads to the relation

$$E_{hAM} = E^\circ_{AM} - \frac{RT}{nF} \ln \frac{a_{Ox}}{a_{Red}}.$$ (2)

The European sign convention conforms closely to the well-established concept of the degree of oxidation of a polyvalent element. In this sense, it would be more correct to say that the Europen convention yields oxidation potentials and the American convention reduction potentials.

---

* The term redox system is used to designate a system in which certain polyvalent ions undergo oxidation or reduction (i.e., lose or gain electrons), depending on the conditions. Examples are $Fe^{\cdots} - Fe^{\cdot\cdot}$, $S^0 - S''$, etc.

Equation (1) is frequently approximated by

$$E_h = E° + \frac{RT}{nF} \ln \frac{[Ox]}{[Red]} ,$$ (3)

[Ox] and [Red] being the respective concentrations of oxidized and reduced forms, expressed in grain-ions per liter.

The necessity of making a correct choice of $E°$ values for Eqs. (1) and (3) must be noted. A constant calculated for the working conditions on the basis of the $Fe^{\cdots}$ and $Fe^{\cdot\cdot}$ concentrations must be used when applying Eq. (3) to the $Fe^{\cdots} - Fe^{\cdot\cdot}$ system, for example. The value for acidic sulfate solutions is $E° = 0.667v$ [2].* The tabulated value of $E° = 0.771$ [1] (the arithmetic mean of $E° = 0.772$ [3] and $E° = 0.770$ [4]) was obtained from calculations based on activities and is to be associated with the $a_{Fe^{\cdots}}/a_{Fe^{\cdot\cdot}}$ ratio of Eq. (1). Introduction of this constant into Eq. (3) will lead to considerable error.

The redox potential is an indicator of the oxidation states of the mineral and organic constituents of a given medium. It is significant that the redox potential of a complex system is fixed by the interaction of all of the various oxidizing and reducing agents that the system contains. The equilibrium value of E_h thus gives an immediate characterization of the medium with respect to all of its polyvalent elements:

$$E_h = E°_{Fe^{\cdots}-Fe^{\cdot\cdot}} + \frac{RT}{F} \ln \frac{a_{Fe^{\cdots}}}{a_{Fe^{\cdot\cdot}}} = E°_{Mn^{\cdots}-Mn^{\cdot\cdot}} + \frac{RT}{2F} \ln \frac{a_{Mn^{\cdots}}}{a_{Mn^{\cdot\cdot}}} = \frac{RT}{2F} \ln \frac{a_{H^{\cdot}}^2}{p_{H_2}} = \ldots \text{ etc.}$$ (4)

Though the redox potential of such systems reflects the over-all state of the medium, it cannot characterize the capacities of the individual redox systems,† or the relative contributions of these systems to the established potential. Measurement of the redox potential must be accompanied by chemical study of the individual redox systems, for this reason.

The redox system contributing most extensively to the over-all redox potential of the medium is said to be potential-determining.

Electrometric measurements with potentiometers and amplifiers are almost always resorted to at the present time for the determination of redox potentials. The inert electrodes are usually made from smooth platinum, the use of gold for this purpose being much less common.

The saturated calomel half-cell is frequently chosen as a reference electrode, its potential with reference to the normal hydrogen electrode being well known.

The value of the redox potential is calculated from the equation:

$$E_{cell} = E_{calomel} - E_h,‡$$

from which it follows that

$$E_h = E_{calomel} - E_{cell}.$$ (5)

An electrode introduced into a medium containing a reversible redox system does not immediately acquire the potential of the latter but builds up to it over a certain interval of time. The rate of establishing this potential depends on various factors, chief amongst which is the redox capacity of the medium. The process of developing the potential consists in the electrode acquiring a definite charge at the expense of the oxidizing or reducing agents which are present in the solution. This charge will be positive or negative, depending on the E_h value of the medium and frequently on its pH as well. It is observed that definite quantities of atomic oxygen or hydrogen are adsorbed here, these gases being formed through electrode discharge of water or hydroxyl

---

*Values of the constant $E°$ based on the concentration ratio, [Ox]/[Red], vary somewhat, depending on the acidity, the nature of the predominant anions, etc.
†The concept of redox capacity is analogous to that of buffer capacity; it is defined by the ratio of oxidized and reduced forms of the polyvalent element in the solution.
‡It is assumed that the calomel half-cell is connected to the "+" binding post of the potentiometer.

and hydrogen ions [5]. Thus the positive charge on platinum is associated with

$$H_2O \rightarrow O_{ads} + 2H\cdot + 2e',$$
$$2OH' \rightarrow O_{ads} + H_2O + 2e',$$

and the negative charge with

$$H + e' \rightarrow H_{ads}.$$

From this it is to be concluded that there is a relation between the electrode potential and the O and H adsorption. An electrode capable of serving as a redox potential indicator will, at the same time, be capable of functioning as a hydrogen or oxygen electrode.

A characteristic potential will be rapidly established on any electrode immersed in a solution containing large quantites of oxidizing or reducing agents (i.e., in a solution of high redox capacity), regardless of the material of which the electrode is constructed, its form, dimensions, and previous history. Different electrodes will, however, show different potential, depending on the material [6, 7] and also on the form and dimensions [8, 9] of the electrode when the solution capacity is quite low. The existence of such a relation is explained by the fact that each electrode has a characteristic capacitance which is fixed by its form and dimensions, and the cleanliness of its surface. L. Michaelis [6] has explained this variation of the measured potential by pointing out that a part of the electrode surface is always occupied by previously adsorbed substances with the amount of material adsorbed and the strength of binding varying from one electrode to another.

Establishment of a potential involves alteration of both the potential of the electrode (at the expense of the redox system of the solution) and the potential of the solution contained in the film surrounding the electrode. Thus a requirement for obtaining trustworthy results is that the capacity of the solution be rather high in comparison with the capacity of the electrode.

The redox potentials of natural substances are found to vary somewhat with the measuring electrode, and it is therefore common practice to take the arithmetical mean of the equilibrium E_h values for 3-5 platinum electrodes as the potential of the medium.

Valid E_h values can be obtained for complex heterogeneous systems such as soils and rocks only after establishment of thermodynamic equilibrium between both the electrode and the solution, and the solid phase minerals of the soil, sludge, or rock. Although colloidal and crystalline substances containing elements of variable valency do not directly participate in establishing the electrode potentials of complex systems, they do serve as a source of ions and thus serve to fix these potentials in the final analysis. The conditions for solution diffusion in heterogeneous systems are poor and the changes produced in the electrode film in such systems are only slowly compensated. Structural and textural inhomogeneities can lead to a variability in diffusional conditions, with E_h the rate of establishing the potential differing from one region to another. These facts make for difficulty in determining redox potentials of heterogeneous systems.

The redox potential of a solution is directly related to its pH and a potential measurement must always be completed by a pH determination for this reason. The measured pH value approximates the negative logarithm of the hydrogen ion activity [10]:

$$pH = - \log a_H. \tag{6}$$

The pH affects the redox potential of a solution by altering the stability of the oxidized and reduced forms of the ions involved in the reversible redox system in different ways. Acid–alkali equilibria are related to oxidation-reduction equilibria through ion hydrolysis, for example. Increasing the pH displaces the equilibria of the $Fe^{\cdots} - Fe^{\cdot\cdot}$ system,

$$Fe^{\cdots} \underset{\nwarrow Fe_2(OH)_2^{\cdots\cdot}\nearrow}{\overset{\nearrow FeOH^{\cdot\cdot} \searrow}{\rightleftharpoons}} Fe(OH)_2^{\cdot} \rightleftharpoons Fe(OH)_3,$$
$$Fe^{\cdot\cdot} \rightleftharpoons FeOH^{\cdot} \rightleftharpoons Fe(OH)_2$$

to the right, but the $Fe^{...}$ concentration falls off much more rapidly than the $Fe^{..}$ concentration and the redox potential calculated from Eq. (3) therefore diminishes.*

The interrelation between redox potential and pH makes it difficult to compare the oxidizing or reducing capacities of solutions which do not have the same pH. M. Clark [1] has proposed that this difficulty be circumvented by the introduction of an index involving both E_h and pH.

Since all redox potentials of a system are the same when measured at equilibrium it follows that any one of the equations of (4) could be used to characterize the oxidation state of a solution. The equation

$$E_h = \frac{RT}{2F} \ln \frac{a_{H^.}^2}{p_{H_2}} \tag{7}$$

is useful since it relates the experimentally determined quantities E_h and pH. Substitution of numerical values for R, T, and F, followed by passage from natural logarithms to logarithms to the base 10, carries Eq. (7) into the form

$$E_h = 0.029 \, (2 \log a_{H^.} - \log p_{H_2}) \, (20^\circ \, C). \tag{8}$$

Log p_{H_2} will be designated by rH (sometimes by rH_2) in analogy with pH which was defined as log $a_{H^.}$, so that:

$$rH = \frac{E_h}{0.029} + 2pH. \tag{9}$$

This index has been used in conjunction with E_h and pH in many studies [12-18, etc.]. It is occasionally employed in place of E_h[19, 20], and has even been referred to as the redox potential [19, 21].

Applications of rH are based on the assumption that unit alteration of the pH of a solution changes the E_h value by 0.058 v (20°C). It is, however, often true that the E_h vs. pH relation for the redox system which fixes the potential of the solution (particularly, a natural solution) is different from that of the $H^.-H_2$ system, so that use of rH is not justified. N. Volk has pointed out, for example, [22] that it is the exception, rather than the rule, that the E_h value of a soil changes by 59 mv (25°C) under unit alteration of the pH; in fact his measurements showed the value of this factor to vary from 58 to 101 mv for various soil types. ZoBell [23] concluded that the $-\Delta E_h/\Delta pH$ ratio is not constant for ocean sludges. W. Pearsall and C. Mortimer [24] have also argued against use of the index rH. The wide favor which the rH enjoyed in the period from 1920 to 1930 has been lost under the impact of such considerations, and it is now significant only in biology where it is still considered a prime measure of aerobic activity [25].

Another method of obtaining comparable results involves the use of a hydrogen electrode of fixed pH, rather than normal hydrogen electrode, for measuring redox potentials. So-called "corrected" potentials of this type are designated by E_{h_2} [26, 27], or $E_h^{'}$ [28, 29]. Redox potentials are sometimes reduced to a single pH [24]. Recalculation of this kind involves the theoretical factor of 58 mv per unit change in pH, and does not, therefore, differ in principle from the use of rH. N. Volk [22] has proposed a most exact method of taking account of the E_h vs. pH relation in which the redox potential is reduced to fixed pH by making use of a $-\Delta E_h/\Delta pH$ correction experimentally evaluated for each case separately.

The natural mobility of a polyvalent element is very closely related to its state of oxidation. Various elements are known in which the ion of higher degree of oxidation hydrolyzes in neutral solution and is therefore unstable. Fe, Mn, Ti, Co, are elements of this type. Such elements tend to concentrate under oxidizing conditions. Other elements, such as U, Mo, and V, have higher mobility and accumulate under reducing conditions.

The fact that the redox conditions are highly significant for the migration, dispersion, and concentration of polyvalent elements in the upper portion of the earth's crust has been pointed out by V. M. Goldschmidt [30], A. E. Fersman [31], L. V. Pustovalov [32], A. G. Betekhtin [33], A. P. Vinogradov [34], B. Mason [35], and others.

---

*The reverse relation is observed in strongly acidic solutions (pH < 2.0) where there is almost no hydrolysis of iron ions.

The role of redox conditions in the formation of deposits of iron [36], manganese [37], light metals [38], and uranium [39] has been treated in various papers.

The disappearance of certain groups of minerals and the formation of others depend on the direction of alteration of the redox potentials and the reaction of the medium. The role of redox potential in paragenesis has been emphasized by V. V. Shcherbina [40]. Calculations have, in fact, shown that minerals containing $V(V)$, $Cr(VI)$, $Mn(IV)$, $Co(III)$, etc. oxidize $Fe^{..}$ ions and cannot exist in the presence of high concentrations of $Fe(II)$ (i.e., in media having low redox potentials) for this reason. $Ti(III)$, $V(III)$, $U(IV)$, and other elements cannot exist at high $Fe(III)$ concentrations.

The redox potential is a determining factor in fixing the native state of an element [35, 41]. Elements can occur free in nature only if their normal potentials are so high that they are reduced by media of ordinary potential and pH.

L. V. Pustovalov [32] has directed attention to the significance of redox processes in the formation of sedimentary rocks. The formation of iron minerals in diagenesis and the dependence of these processes on the redox conditions has been treated in detail in the papers of N. M. Strakhov [42-46], L. A. Gulyaeva [49-49], G. I. Teodorovich [50-52], and A. S. Zaporozhetseva [52].

The redox potential and pH are very significant factors in fixing microorganism activity (i.e., the extent and intensity of biochemical processes) [11, 25], and the direction of transformation of the organic substances which may be present in a deposit. Numerous papers have treated the conditions under which the organic substances present in sludges are converted into petroleum (G. I. Teodorovich [54, 55], L. A. Gulyaeva [47, 49], Z. A. Maimin [56], etc.).

Thus it follows that the redox potential and pH can be important both for prospecting and for interpreting mineral genesis.

The redox conditions prevailing in rocks and waters must be carefully studied in working mineral beds since these factors determine the direction of the water—rock interactions in mine operations. Insufficient attention to the oxidation processes resulting from bed exploitation frequently leads to the appearance of acidic mine waters [57] which extensively damage mine equipment and the surrounding sources of natural water.

The range of problems requiring study of the oxidation-reduction and acid-alkali equilibria characterized by the redox potential and pH is even more extensive than these remarks would indicate.

CHAPTER 2

METHODS OF CHARACTERIZING
THE OXIDATION STATES OF SEDIMENTARY ROCKS

Redox processes are of great importance for geochemistry and this fact has led to numerous attempts at evaluating the oxidation states and redox potentials of rocks.

Study must be made of the redox conditions prevailing during diagenesis of sediments in order to understand the much discussed problem of mineral genesis [18, 42, 47]. Reconstruction of the redox conditions in the geological past requires an embracive study based on present mineralogical compositions and oxidation states. Such studies do not, as a rule, involve determination of redox potentials and pH.

Iron being the polyvalent element most commonly met in rocks, it has become the practice to consider minerals of this element as indicator substances for the study of oxidation states [51, 52, 58]. Sulfur (48, 49,59], and manganese [60] have also been classified as indicator elements. The work of N. N. Strakhov [44, 61] has shown that considerable weight can be attached to the content of residual organic carbon in reconstructing the redox conditions of the geological past.

Redox conditions can be characterized by different methods, depending on the element selected as indicator. The "ground level" oxidation state proposed by V. V. Shcherbina [62] is that prevailing in a medium in which 50% of the iron is reduced, and 50% oxidized. Tables of normal potentials have been drawn on [40] to determine which oxidation states of the various elements can exist with Fe (II) and Fe (III). I. I. Romm [63, 64] has suggested that the oxidation state of a deposit be determined from the Fe (II)/Fe (III) ratio in an acid extract. A. B. Ronov [65] has proposed that the Fe_2O_3/FeO ratio and the organic carbon content be used to characterize the oxidation state of a rock.

The redox conditions prevailing in diagenesis have frequently been reconstructed on the basis of the relation existing between the various mineralogical forms of iron and sulfur [56, 59, 66-74].

L. A. Gulyaeva [47-49] has proposed the development of desulfonation processes as the principal criterion for medium reducibility and suggested that oxidation potentials be determined from the amount of pyrites sulfur, the ratio of pyritic to nonpyritic ferrous iron, and the presence of benthonic fauna. These relations were used in distinguishing four different types of redox conditions, namely: oxidizing, suboxidizing, reducing, and super-reducing.

N. M. Strakhov and E. S. Zalmanzon [44] have shown that the capacity of the reduction processes occurring during diagenesis can be characterized by the amounts of various authigenic iron minerals in the sedimentary rock. Increase in this capacity generally entails a reduction in the redox potential of the medium. These authors therefore proposed to estimate the redox potential of the deposit in diagenesis from the amount of the authigenic iron minerals in the rock, and constructed diagrams for use with sea clays.* N. M. Strakhov claims that similar diagrams could be developed for any type of rock.

These methods lead to fairly trustworthy characterizations of the oxidation states of rocks without giving any quantitative measure of the extent of reduction of one rock as compared with another. Lacking potential

---

* The calculations of W. C. Krumbein and R. M. Garrels [58] were used in this construction.

values, one is forced to describe the redox conditions through such terms as "oxidizing," "weakly reducing," "reducing," "low redox potential," etc. The inadequacy of such characterizations is obvious, especially since classification of a medium as oxidizing or reducing proves to vary with the choice of reference element. Thus L. A. Gulyaeva has shown [47, 49] that the ratio of ferrous to ferric iron is not a satisfactory basis for characterization of the oxidation state of a rock, it being possible for a medium reducing with respect to iron to be oxidizing with respect to an organic substance. G. I. Teodorovich has noted [50, 52] that iron minerals must be used as indicators when characterizing oxidation states with respect to an organic substance, though the reducibility of the organic compound may be one order lower than that of the iron. Thus siderite facies which are weakly reducing with respect to iron may prove to be neutral with respect to an organic substance.

Each mineral is formed and exists under sharply defined physical chemical conditions and becomes unstable and open to disintegration when these conditions no longer prevail. Geochemistry long ago recognized the necessity of determining pH and redox potential existence ranges for the various minerals and relating mineralogical composition with the redox conditions and acidities in various facies.

Various studies have centered around E_h vs. pH diagrams showing limits of existences (i.e., fields of stability) as marked out by theoretical calculations. W. C. Krumbein and R. M. Garrels [58] have constructed diagrams of this kind for iron and manganese minerals and used them as a basis for a system of deposit classification. The iron diagram was subsequently used by H. L. James [36] to explain the origin of various facies. Similar diagrams for copper, zinc, and lead have been developed by R. M. Garrels [39]. Variant diagrams showing the fields of stability of iron and manganese were used by K. Krauskopf [37] to explain the separation of iron and manganese under natural conditions.

Each of these determinations of the stability fields of the various minerals were based on formal calculations and various conventions and the E_h and pH values obtained are only approximations. Thus the existence limits of the sulfides were determined from the $SO_4^= - S^=$ redox potential although this system is never met in practice. N. K. Huber and R. M. Garrels [75] have experimentally tested the iron diagram and found it correct in general outline, but even they admit that such diagrams stand in need of practical check and correction and can be used only for purposes of orientation. Clear cut relations cannot be established between the paragenetic mineral association and the physical chemical conditions prevailing in rocks without direct determinations of redox potentials and pH. The methods currently used for the determination of redox potentials of sedimentary rocks do not, however, lead to trustworthy results.

Redox potentials are now being used to characterize oxidation states, not only in physical and analytical chemistry, but in various other sciences as well. Scores of papers have been devoted to problems in soil science, to the study of the redox conditions prevailing in various soil types and their dynamics, to the effect of various factors on the redox potentials of soils, to the effect of redox conditions on productivity, and so on [12, 15, 76-85, etc.]. Redox potential measurements are frequently carried out in the study of existing ocean [23, 28, 29, 86-94] and lake [17, 20, 95-97] sludges. Study has been made of the redox potentials of natural waters [98-106], bacterial cultures [9, 25, 27, 107], biological liquids and tissues [7, 108], plants [16, 109-111], food stuffs [112], sewer water [113], and caustobioliths [114].

On the other hand, we know of only three papers on methods of measuring redox potentials in sedimentary rocks [115-117] and this despite the fact that these potentials have been drawn on in innumerable works to explain the various rock processes.

The method of Itkina [115] involves placing the powdered rock in a measuring flask where it is moistened and carefully mixed.* The flask is then closed with a stopper carrying platinum and calomel electrodes and flushed out with a current of nitrogen or carbon dioxide. The platinum electrodes are to be previously treated for two hours with a chromate mixture. Determination of the redox potential requires 2-3 h.

This method leads to correct results only with highly oxidized rocks. It must be remembered that the redox potential is associated with the liquid phase. In measuring the potential of the rock, one is, in actuality,

---

*E. S. Itkina found that the best results were obtained at a moisture content equal to 70% of the moisture capacity of the rock, i.e., with an approximate 2:1 rock to water ratio.

measuring the potential of a mineral solution. The measured value represents the potential corresponding to the oxidation state only if the rock impregnating solution is in equilibrium with the rock minerals. The Itkina method requires air grinding of the rock and addition of water. This leads to extensive oxidation of the mineral solution. Subsequent flushing with nitrogen can serve no essential purpose. The fact that air can affect the measured redox potential is indicated by the work of Itkina herself, where the potentials of certain silts were reduced by 150 mv when these operations were carried out in an atmosphere of nitrogen. Use of CO_2 as an "inert" gas reduces the pH and thereby increases the value of E_h. This effect has been observed by M. S. Zakhar'evskii [112] in redox potential measurements on meat products. Preliminary treatment of the platinum electrodes with a chromate solution sharply increases the amount of oxygen adsorbed by the metal and thus leads to a rise in the characteristic redox capacity. It will be shown below that the second reaction in the following scheme is the slow step in establishing a redox potential:

$$\text{Electrode} \rightleftharpoons \text{Solution} \rightleftharpoons \text{Rock minerals}.$$

A period of 2–3 h is entirely inadequate for establishing even the first of these equilibria. It is no accident that there are no values less than +297 mv among the great number of measurements cited by E. S. Itkina, most of which fall in the interval from +350 to +450 mv, i.e., in the interval corresponding to oxidized rocks.

The following method has been recommended in a paper by L. V. Pustovalov and E. I. Sokolova [116]. A piece of rock weighing 40–50 g is pulverized in an agate mortar. The resulting powder is placed in a glass tube closed at one end with gauze and filter paper. The tube with the powdered rock is then placed on a stand in a desiccator containing water which has been thoroughly boiled to remove oxygen and carbon dioxide. The gauze enclosed end of the tube is immersed in this water. The powder in the tube becomes saturated with water in the course of 18–24 h and is then transferred to a weighing bottle whose cover carries platinum and calomel electrodes. pH measurements are carried out by replacing the platinum electrodes with antimony or glass electrodes. Measurements of the redox potential require from 40 min to 2 h. The authors recommend that measurements be concluded when the potential alters by less than 2 mv in 5 min. It is suggested that the platinum electrodes be first treated with either a chromate solution or a 10% HCl solution, and then washed with alcohol or distilled water.

This method is open to the same objections as were advanced against the Itkina procedure, namely that it cannot assure trustworthy results in the case of reduced rocks, since no attempt is made to prevent oxidation in the preliminary handling and in the course of the measurements themselves. The E_h vs. pH graph presented by these authors shows no E_h value less than +100 mv among the numerous measurements, and this despite the fact that some of the mineral samples contained siderite, a substance which forms and exists under reducing conditions. It is noted that the results obtained here depart considerably from those calculated from the Krumbein and Garrels diagrams [58], but this is explained in terms of inaccuracies in the diagrams themselves.

These same defects are also inherent in the methods outlined in the manuscript instructions of E.I. Sokolova, L. P. Listova, and A. Z. Vainshtein [117]. These authors have tested methods of measuring E_h values for rocks and minerals, in suspension and as moist powders, and have also attempted direct measurements on loose rocks in place. In none of these procedures was adequate attention given to the possibility of rock oxidation, during preparation and in the course of the measurements. Even the authors admit that the results obtained were not trustworthy. Negative values of E_h were obtained in numerous instances, but these drifted and were poorly reproducible. Steady state values of the potential were established only slowly.

Unsatisfactory results were also obtained in redox potential measurements on sedimentary rocks which were undertaken by T. I. Kazmina, Z. S. Gerasyuto, T. S. Rogachevskaya [118], and by N. N. Yurganov [73],* the measured values being higher than anticipated from the known rock compositions. Thus none of the E_h values obtained by N. N. Yurganov in his measurements on various rocks was lower than +100 mv. For this reason, many of the E_h values were calculated (72–74) from Eq. (3), using the value of the Fe(III)/Fe(II) ratio for a 5% HCl extract and introducing a pH correction.† The assumption was made in [73] that the ratio of the iron forms in

*The literature does not clearly indicate the methods followed in this work.

†The papers of N. N. Yurganov do not indicate how this correction was made.

the extract remains constant at the value existing in the deposit during diagenesis. This assumption is, however, erroneous, if for no other reason than that pyrite – one of the principal factors in fixing the oxidation state of the rock – is completely insoluble in acids and thus can have no effect on the E_h value. Our own data would indicate that the lowest values of the Fe(III)/Fe(II) ratio for acid extracts are met with rocks having high siderite content (regardless of the E_h value), whereas very low redox potentials are characteristic of rocks containing pyrites.

Examples of the calculation of E_h from the ratios of the various forms of iron, sulfur, and manganese in acid extracts have been given by V. G. Savich [91] for existing sea deposits. Comparison of the measured redox potentials with values calculated from the equilibria

$$Fe''' \rightleftarrows Fe'', \quad S^0 \rightleftarrows S'' \quad \text{and} \quad Mn'''' \rightleftarrows Mn'',$$

made it possible to mark out the potential-determining systems in certain deposits. This work indicated that good agreement between the calculated and measured redox potentials is achieved only rarely. The papers of N. N. Yurganov [72-74] report quite a few redox potential values of the order −400 mv (up to −461 mv), although such values are not met, under even the most extreme reducing conditions in existing deposits, if exception is made of those described by K. ZoBell [23].

The method proposed by I. Yuranek [120, 121) for the measurement of E_h in acid extracts is not essentially different from that suggested by N. N. Yurganov.

D. Bardossy and M. Bod [122] have suggested that experimental determination of redox potentials in sedimentary rocks is impossible in principle since the added distilled water is a poor solvent and does not dissolve the various redox components of the rock to the same extent. For this reason, it was proposed [122, 123] that rock comparisons be limited to reduction capacities.

Failure of redox potential measurements on sedimentary rocks is certainly responsible for the appearance of statements to the effect that such measurements have no significance for ancient rocks [56, 70, 71], the results obtained reflecting the present state rather than the character of the environment during diagenesis [56, 61]. On the other hand, many authors in addition to those cited above have reconstructed past geochemical conditions from the authigenic mineral content of rocks. This assumes that sedimentary rock compositions are either constant or vary only slightly. The law of physical chemical succession of L. V. Pustovalov [32, p. 396] would indicate, that the physical chemical conditions in rocks which maintain a definite complex of authigenic minerals are the same (or approximately the same) as those prevailing at formation. L. V. Pustovalov and E. S. Sokolova write [116, pp. 124, 125]: "There is every reason to believe that just as the physical chemical characteristics of the present state of a rock reflect the chracteristics of the initial deposit, so does the present chemical−mineralogical composition of the rock reflect the composition of the initial deposit." "The sedimentary rock brings not only syngenetic minerals from the distant geological past but also a reflection of the physical chemical conditions prevailing in the period of formation, conditions which we are not yet, unfortunately, able to decipher." These authors point out that the methods of measuring E_h and pH for rocks are undeveloped and the application of these values in geology questionable, despite much work in this field. No trust can be put in the E_h and pH values obtained for rocks. N. N. Yuragonov has expressed the opinion [18] that values of the redox potentials of sedimentary rocks obtained with the electrometric method are more nearly correct than values obtained by indirection, and this despite the lack of trustworthy procedures for ancient rocks.

It is clear that insufficient attention to the conditions of handling test samples of rock has frequently added to the difficulties of measuring redox potentials. K. F. Rodionova, E. V. Podol'skaya, and A. I. Volodchenkova [71] have referred to the work of L. V. Gulyaeva, E. S. Itkina, and T. I. Kazmina in pointing out that the time elapsing between sampling and testing is an important factor in determining the measured value of E_h.

Thus there is a need for a direct method of determining redox potentials in sedimentary rocks which would lead to values more accurately characterizing the oxidation state. The various factors making for difficulty in such measurements will be considered below.

CHAPTER 3

DISTINGUISHING FEATURES OF REDOX POTENTIAL MEASUREMENTS IN SEDIMENTARY ROCKS. APPARATUS FOR MEASUREMENTS IN AN ATMOSPHERE OF NITROGEN

The methods employed for measuring redox potentials in materials such as soils and sludges must be radically altered to allow for the peculiarities of sedimentary rocks. Most rocks must be pulverized, since they are naturally harder (i.e., more highly consolidated) than soils and sludges and electrodes cannot be introduced into them. Moreover, water must frequently be added to the rock, the moisture content of the latter being too low to assure the necessary degree of contact between the impregnating solution and the electrode.

The principal difference between a rock, on the one hand, and a soil or sludge, on the other, is that the mineral solution* systems of the rock have vanishingly low redox capacities, even though the rock itself is an abundant source of potential determining minerals. This conclusion is supported by the fact that the time required for establishing a redox potential is considerably longer for rocks than for any other type of material.

This difference probably traces back to the fact that there is a marked reduction in polyvalent element mobility during the rock's extended history, less stable compounds being converted into compounds of higher stability, the less soluble colloidal state being replaced by the crystalline condition, and so on. Thus the data of N. P. Tsyb [124] on rocks taken from the vicinity of the Volgograd hydroelectric station indicate that the mineral solutions of reduced rocks do not contain iron, even though this is the potential determining element.

The redox capacity of the rock system of impregnating solutions is low and the redox potential can therefore be markedly affected by comparatively insignificant external factors such as oxidation by atmospheric oxygen. Accordingly, the method adopted for measuring redox potentials of rocks must provide for careful isolation of the rock specimen from the air from the time of collection to the completion of measurements, for grinding and moistening in an atmosphere of nitrogen, and for prior removal of dissolved water-oxygen from the water. The low capacities of the redox systems also require selection of a preliminary treatment which will reduce the characteristic capacity of the electrodes [8, 9, 107].

These conclusions were reached only after our methods of measuring redox potentials had passed through a number of stages. These methods were developed with cores taken from borings in the vicinity of the Volgograd hydroelectric station and from rock samples from excavations for the foundations of the station. Redox potentials were measured with a PPTV-1 potentiometer with a Kryukov amplifier [125, 126]. A portable lamp-and-scale potentiometer was used for field measurements [126].

Platinum electrodes of the type used here are shown in Fig. 1. Electrode a was prepared by sealing a platinum wire, 1 mm in diameter, into the drawn out end of a glass tube, the length of the exposed portion of the wire being approximately 1 cm. An electrode of this kind was simple to prepare and sturdy enough to be directly introduced into many test rock samples in the natural state, and such electrodes were therefore used more

---

*These remarks are principally directed to reduced rocks, potentials usually being rapidly established in oxidized rocks.

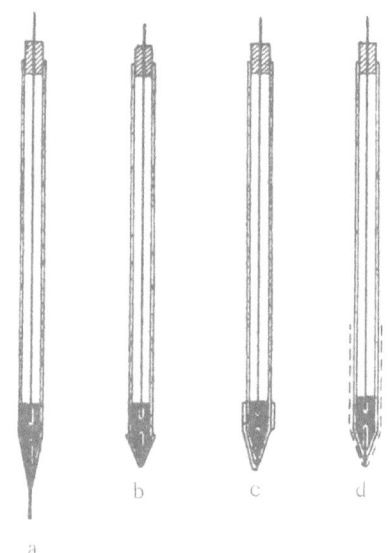

Fig. 1. Types of platinum electrodes.

often than any other type. Electrodes b and c were prepared from platinum foil, the one having the form of a cone, and the other that of a cylinder, fused to the end of a pointed glass tube. Electrode d was prepared by covering the end of a glass tube with a thin layer of platinum, this layer being obtained by weak thermal reduction of platinum compounds [127, 128]. A thin platinum wire was used to make contact between the platinum covering of electrodes b, c, and d and the mercury inside of the electrode. A saturated calomel half-cell of the type described by P. A. Kryukov [125, 126, 129] was used as a comparison element.

The first determination of redox potentials were carried out directly on paraffined specimens (cores). A hole somewhat smaller in diameter than the electrode tube was made in the protective layer of gauze and paraffin with a borer, the upper layer of rock was removed through this opening, and an electrode introduced, the depth of penetration being 2-10 cm, depending on the hardness of the sample. The position entry of the electrode into the core was then sealed over with molten paraffin. This same method was also used in the first field potential measurements on rock outcroppings from the foundation slopes of the Volgograd hydroelectric station, wet clay being employed in place of paraffin.

This method did not give acceptable results when applied to reduced rocks; the readings obtained with the various electrodes were different, poorly reproducible, and inconsistent with the values to be theoretically expected from the known mineralogical composition of the rock. Positive E_h values were usually obtained, regardless of the rock under study.

This situation largely resulted from the entry of air into the cleavages made by introduction of the electrode. This led to the formation of an oxidized zone in the portion of the rock adjacent to the electrode where the redox potential of the rock impregnating solution was increased by passage from reduced to oxidized forms in the reversible redox systems. Lacking an appreciable solution exchange in the neighborhood of the electrode, diffusion and chemical reaction with the rock minerals gradually led the solution back into its original state as the potential was being established. An instance of this kind is found in

$$2Fe''' + FeS_2 \rightarrow 3Fe'' + 2S^0.$$

Since the capacities of the reversible redox systems of mineral solutions and the rock moisture content are both low, such oxidized regions around an electrode can remain unreduced for extended periods. The fact that atmospheric oxygen is a determining factor in fixing the redox potentials of rocks means that the Vainbaum method [130, 131] (recommended for E_h measurements in gas wells) cannot be employed here, this method requiring that the platinum be placed on the surface, rather than within the rock.

These facts suggested the necessity of carefully protecting the rock from the atmosphere during grinding, moistening, and measurements. The aim of grinding and moistening is to improve the contact between electrode and rock, and between the portion of rock in contact with the electrode and the remainder of the sample. Grinding also makes it possible to introduce the electrode into the rock, the hardness being such that this frequently cannot be done in the natural state.

Simple introduction of the electrode into the natural rock is often entirely adequate, especially with oxidized rocks and reduced plastic rocks of relatively high moisture content (35-45 %).*

It was found that a steady and reproducible redox potential is established in the first hour of measurements on oxidized rocks. Here the rock does not need to be carefully protected from the air, and frequently does not

___
*Ancient loams and clays which are still clearly in diagenesis and contain hydrotroilites with considerable quantities of organic matter.

20

TABLE 1. Redox Potential Measurements on a Sandy Siltstone from the Tsartsyn Deposits

Time from beginning of measurements	E_1, mv	E_2, mv	E_3, mv	E_4, mv	E_5, mv	E, mean of readings on 5 electrodes
15 min	573	564	579	578	574	574
30 min	579	572	584	584	575	579
1 hr 30 min	582	577	591	591	580	584
2 hr 00 min	586	578	593	593	580	586
3 hr 00 min	586	577	592	592	578	585
4 hr 00 min	587	576	594	593	580	586
5 hr 00 min	587	576	594	593	581	586
23 hr 00 min	591	578	600	595	584	590

even require preliminary moistening. The results of certain core redox potential measurements are presented in Table 1 to illustrate the stability of the values obtained with oxidized rocks.*

The fact that a steady potential was established rather rapidly in these oxidized rocks was due to the liberation of free sulfuric acid during the oxidation of pyrites. This markedly increases the mobility of iron and leads to a rise in the redox capacities of the mineral solutions. While concentration of iron in the mineral solutions of reduced rocks is so low as to defy analytical detection of the element, its value can be as high as several grams per liter after oxidation [132]. Moreover, equilibrium is reached quickly in a medium of high E_h, the potential originally registered on a platinum electrode differing but little from the potential of the medium itself.

It is therefore most convenient to measure the redox potential of an oxidized rock in an open vessel, working with the rock pulverized and slightly moistened (to the consistency of a thick paste), and immersing the electrodes to the bottom of the vessel.

Isolation from atmospheric oxygen and good electrode—rock contact resulting from high plasticity at high moisture content make it possible to use this same method in measuring redox potentials of reduced plastic rocks (in distinction to other reduced rocks).†

Since direct E_h measurements on natural rocks did not generally lead to correct results, a number of devices were constructed for making such measurements in an atmosphere of nitrogen.

Apparatus No. 1 was designed for measurements of redox potentials in cores [133]; it is shown in Fig. 2. The principal part of this device was a plexiglass cylinder, 1, with a freely rotating plexiglass cover, 2. This cover carried six ground joints, 3, located at the vertices of a regular hexagon concentric with the cover itself. Three platinum electrodes, 4, and a calomel electrode‡ were inserted into four of these joints; the fifth carried a borer‡ for making an opening in the core sheathing and the sixth a drill, 5, for loosening the rock. The center of the cover carried a tube, 6, for the admission of oxygen-free water. Two tubes, 7, on the sides of the cylinder served for the introduction and removal of nitrogen.

Prior to measurement, the apparatus was set on the top of the core and a mixture of molten paraffin and asphalt poured around the base-core joint. The apparatus was flushed out with nitrogen and four openings made in the core surface, 9, by the borer, these openings being at 60° from the center of the apparatus. The rock in these openings was then loosened and moistened with the deareated water, and the electrodes introduced into

*These cores were taken from a layer of rock lying above the oxygen zone.
†Preliminary electrode polarization was used to show that the redox capacities of the impregnating solution systems are considerably higher in ancient loams than in other types of reduced rocks. The time required for electrode depolarization in these rocks is measured in hours, instead of days as with other reduced rocks.
‡Not shown in the figure.

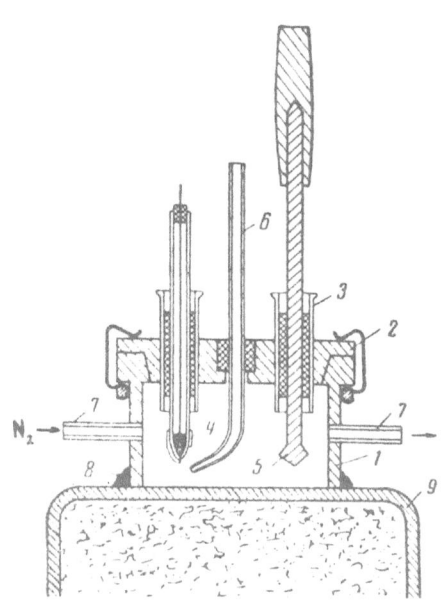

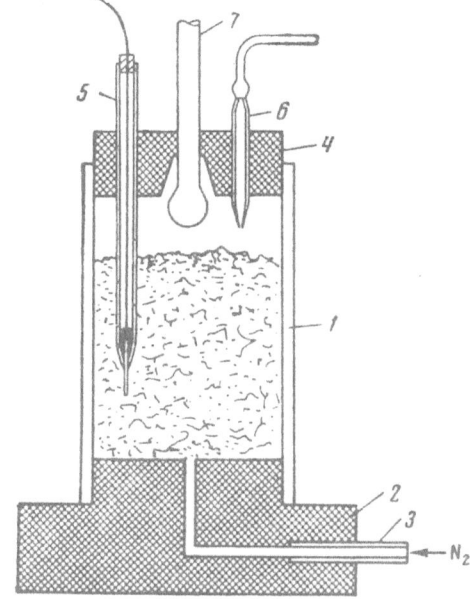

Fig. 2. Apparatus No. 1. Fig. 3. Apparatus No. 2.

the resultant paste. Nitrogen was passed through the apparatus during these preliminaries and at the beginning of the measurements themselves. Entrance of air into the system was at the same time avoided by connecting the nitrogen inlet tube to a Tischenko flask filled with a $Fe(OH)_2$ suspension. The difference of the liquid levels in this flask made it possible to check the sealing of the apparatus in cases where the measurements could not be concluded in one day or the current of nitrogen failed. There was difficulty in cutting the openings in thick-covered cores; here the covering and underlying rock were cut away and rapidly paraffined over before setting the apparatus in place.

This apparatus made it possible to have perfect protection from the atmosphere during measurements; its weakness lay in the low mechanical strength of binding to the core.

Apparatus No. 2 was more simple in construction (Fig. 3); it consisted of a thick-walled glass cylinder, 1, and two rubber stoppers, 2 and 4. The lower stopper, 2, served as the base for the apparatus and carried a tube, 3, for the introduction of nitrogen. The upper stopper, 4, carried three platinum electrodes, 5, a calomel electrode,* a tube, 6, for the introduction of water and another tube for the introduction of nitrogen,* the latter being connected to a Tishchenko flask containing a $Fe(OH)_2$ suspension. A glass pestle, 7, was set in the center of this stopper.

The test rock cut in cylindrical form was set inside the apparatus and the latter closed by the stopper carrying the electrodes. Air was then displaced from the apparatus, oxygen-free water intoruced, the rock pulverized by the pestle, and the electrodes lowered into the resulting paste.

This apparatus was convenient for field measurements since only an insignificant quantity of nitrogen was required for displacing the small volume of air remaining after introduction of the rock.

Apparatus No. 3 [134] (Fig. 4) was so constructed that the pulverized rock could be agitated in the course of the measurements, thus considerably shortening the time required for establishing equilibrium with the electrodes.

This apparatus was built around two thick walled glass jars one set inside the other, the innermost, 1, being joined to the outermost, 2, by a rubber ring, 3. A glass cylinder, 4, somewhat larger in diameter than the

---

* Not shown in the figure.

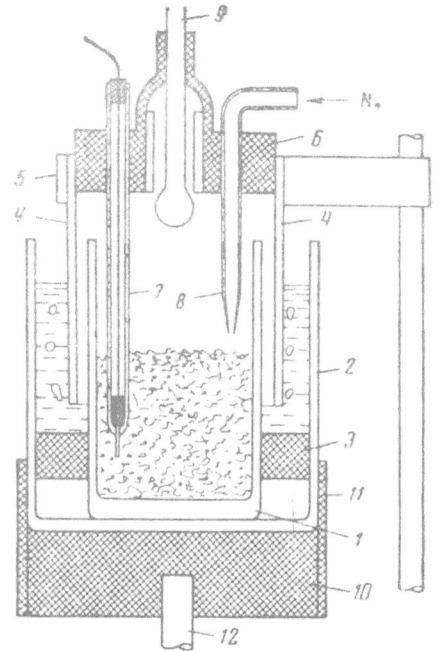

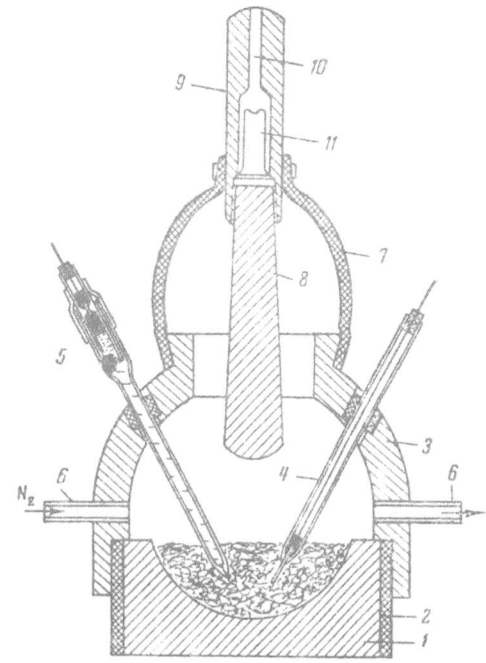

Fig. 4. Apparatus No. 3.   Fig. 5. Apparatus No. 4.

inner jar, was inserted between the jars and held in position by a clamp, 5. This cylinder was closed at the top by a rubber stopper, 6, carrying three platinum electrodes, 7, a calomel electrode,* a tube for the introduction of deareated water,* and a tube for the admission of flushing nitrogen,* 8. A glass pestle, 9, was inserted in the center of this stopper. The jars were attached to the axel, 12, of a motor decelerator by a rubber stopper, 10, and ring, 11. A saturated NaCl solution containing a small amount of $TiCl_3$ was introduced into the space between the jars prior to measurement of the redox potential. A small cylinder of the test rock was inserted into the inner jar. The apparatus was then closed with stopper, 6, carrying the electrodes, and the apparatus flushed with nitrogen, the gas passing out through the solution seal between the jars and cylinder, 4. The rock was pulverized by the pestle after complete displacement of air, water was added, and the electrodes lowered into the resultant paste.

The jars were caused to rotate at 20-30 rpm, thus agitating the rock while the electrodes are held firmly in place. Measurement of the redox potential was made 15 min after the completion of agitation. This device suffered from the same defect as the preceding, namely that it could be used only with soft rocks.

Apparatus No. 4 (Fig. 5) was designed by P. A. Kryukov [133] to permit pulverization of harder rocks. It consisted of an agate mortar, 1, set into a rubber ring, 2, and a massive plexiglass housing, 3. Rubber and chlorvinyl packing was used to affix four platinum electrodes, 4, a calomel electrode, 5, two tubes for the introduction and removal of nitrogen, 6, and a tube for the addition of water* into this housing. The nitrogen inlet tube was connected to a Tischenko flask containing a $Fe(OH)_2$ suspension. The top of the housing was closed by a rubber bellows, 7, carrying an agate pestle, 8, with handle, 9. In order that air be removed from the upper portion of the apparatus, a canal, 10, was bored through the pestle handle, and a one-way outlet valve, 11, introduced into it.

The various operations with this apparatus were carried out in the following order. A 100-120 g sample of rock from the core was surface washed and inserted into the mortar, the cover with attached electrodes set into place, and the apparatus flushed out with nitrogen. The electrodes were arranged in such a manner as to

---

* Not shown in the figure.

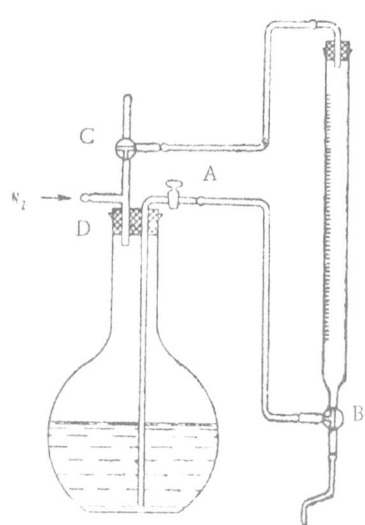

Fig. 6. Vessel for the preparation
and storage of oxygen-free water.

offer no interference to the movement of the pestle. The opening in the pestle handle was closed at the completion of the nitrogen flushing, water added, the rock crushed by the pestle, and the electrodes lowered into the resultant paste.

This was the most generally applicable and trustworthy of the various devices, and most of our measurements of rock redox potentials were carried out with it.

It was observed that the potential of the calomel electrode sometimes altered in the course of extended E_h measurements. For this reason, the working calomel electrodes were always checked at the end of measurements, check being against a calomel electrode which had not been used in the measurements.* The saturated KCl solution in the electrode tube was replaced and a correction introduced into the E_h calculation if the check showed a departure of more than 5 mv. The calomel electrodes were stored in a saturated KCl solution.

Nitrogen was prepared by passing air through a $Fe(OH)_2$ suspension and over copper heated to 500°C, final purification being by passage through a $CrCl_2$ solution. The nitrogen current was so regulated that two bubbles per minute passed through the Tishchenko tube with the $CrCl_2$ solution during the preliminaries to the measurements, and one per second at the beginning of the measurements themselves. Rubber balloons filled with nitrogen in the laboratory were used in the field measurements.

Water was freed of oxygen and stored in the vessel shown in Fig. 6. The oxygen was removed by boiling distilled water for a half hour, nitrogen being passed through stopcock A during the last 10 min of the heating. The water was slowly cooled and subsequently used with the nitrogen supply directly connected to tube D. The burette and cocks C and B could be used for measuring definite volumes of water without admitting air to the system. By using this apparatus and carrying out the measurements in an atmosphere of nitrogen, it was possible to obtain trustworthy and reproducible redox potentials which were consistent with the known chemical and mineralogical analyses of the rock [119, 132].

Use of Apparatus No. 2, 3, or 4 presumed that the opening of the core and transfer of the sample would be carried out in the air. It is to be remembered that oxidation affects only the surface solutions on the rock, the oxidized surface of the test piece making up only an infinitesimal part of the entire grain mass resulting from pulverization in nitrogen.

Nevertheless, the fact that the sample was not protected from the atmospheric oxygen during transfer to the apparatus is one of the defects of this method. This fact must be taken into account in attempting further improvements in the apparatus and methods of measuring redox potentials of rocks.

---

*It is necessary to periodically check the calomel electrodes against a hydrogen electrode.

CHAPTER 4

THE EFFECT OF MOISTURE

It has been noted in the preceding chapter that a sedimentary rock must be not only protected from atmospheric oxygen but moistened as well if its redox potential is to be determined. In fact, it is impossible to establish electrochemical equilibrium between rock and electrodes unless there is a satisfactory contact between electrodes and impregnating solution.

The rate of establishing the electrode-solution equilibrium is largely determined by the ratio of the redox capacity of the mineral solution system and the characteristic redox capacity of the platinum electrodes.* The potential of a reduced rock initially registered by a platinum electrode* which has been in previous contact with the air is quite different from the E_h of the mineral solutions, and the redox capacity of the solution system is low. From this it follows that equilibrium will be established slowly, even if the contact between the electrodes and the rock impregnating solution is good. Equilibrium may be established with infinite slowness in natural sedimentary rock since the characteristic moisture contents are too low to ensure good contact. This same point has been made by L. V. Pustovalov and E. I. Sokolova [116] who have concluded that moistening is the only method now available for obtaining trustworthy results in redox potential measurements on rocks.

The effect of moistening is illustrated in Fig. 7 where electrode potentials obtained on moistened and unmoistened rock are compared. The rock samples were ground under nitrogen in one of the previously described pieces of apparatus and the electrodes immersed in the resulting mass. Deaerated water was introduced several hours after beginning measurements. Each experiment showed that there was a pronounced reduction of the potential as soon as the rock was moistened.

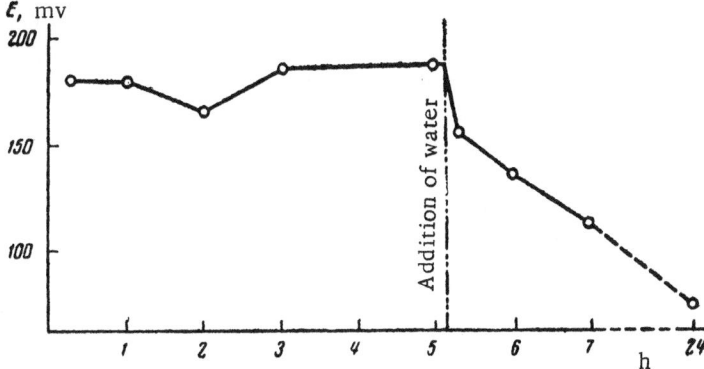

Fig. 7. The effect of moistening on the electrode potential vs. time curve obtained in E_h measurements on a sandy siltstone of the upper Tsaritsyn deposits.

---

*The meaning of these terms will be explained in Chapter 6.

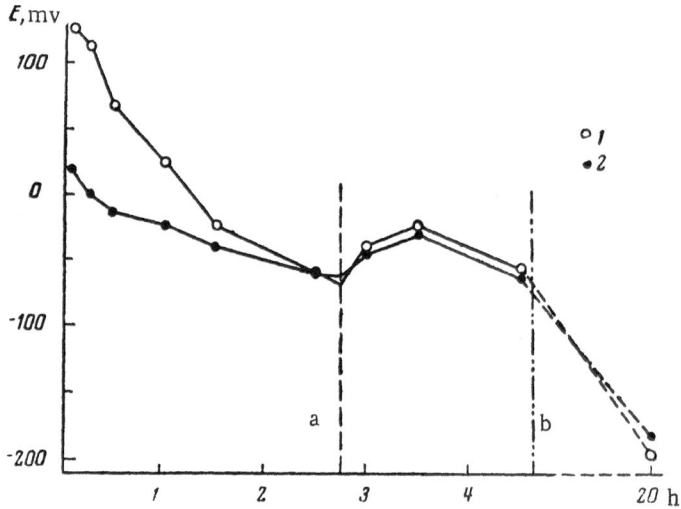

Fig. 8. Alteration of the electrode potentials during E_h measurements on a sample of sandy siltstone from the upper Tsaritsyn deposits with rock agitation during measurement (rock, 80 g; water, 10 ml): 1) electrode treated with chromate solution; 2) electrode treated with NaOH solution; a) 10 ml additional water added; b) agitation completed.

Cases were observed in which the potential increased up to the instant of introducing the water, such effect being quite common with electrodes which had been in contact with the atmosphere for a considerable time.

The addition of water affects the establishment of the potential in diametrically opposite ways. Moistening gives that improvement of contact between electrode and rock, and between the various rock sectors, which is required for obtaining correct results, but the addition of water breaks down the equilibrium existing between the rock and the mineral solutions, and an extended period of time is required before this equilibrium can be reestablished.

The alteration of the redox state of the mineral solutions is not solely due to dilution; other factors affecting it are: partial oxidation during removal of the rock sample from the core; the introduction of oxygen by the electrodes; the presence of oxygen in the apparatus at the beginning of the nitrogen flushing, and so on. The combined effect of these various factors is to raise the redox potential to a positive value. The potential gradually falls under interaction of the solution with the rock minerals, and approaches the value characteristic of the mineralogical composition. The rate of approach is here dependent on the extent to which the mineral solution has been altered in the preliminary operations and on the absolute content and states (degree of dispersion, for example) of the minerals which are formed under reduction and support the reducing conditions by their very presence.*

The fact that redox potential measurements on sedimentary rocks extend over long intervals of time is indication that equilibrium between solution and rock minerals is established slowly here. This same conclusion is also indicated by the results of experiments with Apparatus No. 3 which permitted agitation of the rock during measurement. Figure 8 shows the results obtained in an experiment in which one of the electrodes was given a prior treatment with a chromate solution and the other treated with NaOH. It is obvious that the initial oxidation states of the electrodes were not the same, the potentials recorded becoming equal only three hours after

---

*A specific cause for the reduction of mineral solution redox potentials can be found in the accumulation of ferrous ions formed by reaction between pyrite and ferric ions.

TABLE 2. Results of Redox Potential Measurements with Various Degrees of Moisture on a Sample of Sandy Siltstone from the Upper Tsaritsyn Deposits

Exp. No.	Rock taken, g	Added water, ml	Weight ratio, rock to added water	E_h, mean value for 3 electrodes, mv	Remarks
1	185	18.5	10 : 1	−224	Rock in the form of a thick paste
2	140	20	7 : 1	−190	Unstratified paste
3	160	40	4 : 1	−231	Poorly stratified suspension
4	80	80	1 : 1	−224	Markedly stratified suspension

the beginning of measurements. It is to be concluded that the electrodes have then come into equilibrium with the solution, further temporary reductions in the potential being ascribed to an alteration of the oxidation state of the solution. The increase in potential resulting from the second moistening was probably due to the presence of traces of oxygen in the added water. Equilibrium between the platinum electrodes and the solution is established much more slowly when the rock is not agitated during measurement, 2-5 days being required, but the establishment of the solution−rock mineral equilibrium is the slow step, even then.

It is clear that the amount of added water should be held to a minimum to reduce the undesirable effects resulting from moistening. It is recommended that the pulverized rock be moistened just to a thick paste, this state corresponding to a rock to added water ratio of 10 : 1 to 15 : 1 in the case of the rocks which we have studied. Increase in the amount of added water usually increased the time required for the measurements. The results obtained will, however, be independent of the degree of moistening if the measurements are extended over a long period. Table 2 shows the results of parallel measurements on a single sample of rock, these measurements being carried out under different degrees of moisture in four different systems of type No. 2. A common final value of the redox potential was obtained in measurements over two months despite the fact that the rock to added water ratio was varied by a factor of ten in these experiments.

It has been repeatedly noted that the development of reduction processes leads to a diminution in the redox potential of a rock which has been moistened and protected from contact with the air [13, 15, 78, 135]. Similar effects are observed in air-protected sludges [64]. This is to be explained in terms of an aerobic bacterial reduction of the reversible redox systems of the rock or sludge by organic matter. Extended reduction of redox potentials in sedimentary rocks cannot be accounted for in this way since:

TABLE 3. Determination of the Redox Potential of a Sample of Sandy Siltstone of the Upper Buchak Deposits (Measurements Carried out in an Atmosphere of Nitrogen and with Moistening)

Time measured from the beginning of measurements	E_1, mv	E_2, mv	E_3, mv	E_4, mv	E, mean of measurements on 4 electrodes, mv
2 hr	703	703	650	712	692
6 hr	707	701	656	723	697
1 day	703	682	653	718	689
3 days	708	677	655	702	686
4 days	705	677	657	701	685
5 days	707	678	661	702	687
11 days	699	670	666	705	685
19 days	695	667	658	701	680

1. Reduction processes are not to be expected in rocks which have existed under anaerobic conditions in the course of geological time, since the active organic ingredients would have been consumed in diagenesis.

2. The final potential values are stable, reproducible, and in good agreement with the E_h values anticipated on the basis of theoretical calculations.

3. Moistening and protection from the air do not lead to a reduction of the E_h values obtained for strongly oxidized rocks,* even though the latter contain essentially the same amount of residual organic material as rocks which were formed under reducing conditions. The results presented in Table 3 illustrate this point.

These facts indicate that the oxidation state of the rock rather than accidental processes will fix the E_h value reached after a long sustained potential drop.

The moistening preliminary to measurement is one of the principal causes of the long time required for the establishment of a potential and further improvement must therefore look to the development of a technique for making measurements at the natural moisture content of the rock.

It is clear that measurements under such conditions would have to be carried out with low-inertia electrodes,† or with specially treated platinum electrodes of the usual type, in apparatus which would permit agitation of the pulverized rock.

---

*Provided that these rocks do not contain large amounts of pyrite.

†The meaning of this term will be considered below.

CHAPTER 5

THE TIME REQUIRED FOR MEASURING
REDOX POTENTIALS OF SEDIMENTARY ROCKS

The time required for measuring redox potentials of reduced sedimentary rocks is considerably greater than that required for E_h measurements on other types of natural products.* Thus the maximum time required for establishing a redox potential is 4 h in measurements on natural waters [99], 2-3 h in measurements on sea deposits [17], and 15-18 h in measurements on soils [12, 136]. On the other hand, it is frequently observed that the potential of a reduced rock continues to fall for 7-15 days.

Since approach to the redox potential of the rock is always from the positive side and accidental factors invariably increase the potential, the lowest of the mean values of the readings on the various electrodes† was taken as the rock potential, even if this was not the same as the final reading.

The fact that extended periods of time are required for the establishment of a potential in reduced rocks traces back to the low redox capacity of the mineral solution system and the alteration in the oxidation states of the solutions of this system during preliminary treatment.

Redox potential measurements on various natural products are frequently concluded when the alteration of potential is less than several millivolts per hour. It has been recommended, however, that such measurements be stopped when there is no more than a 1-2 mv alteration in potential in the course of 5 min [116, 137]. Although short term measurements of this kind are entirely adequate for soils in which the redox equilibrium is easily displaced, they are quite unacceptable in the case of rocks. The diminution of the electrode potential is here so slow that a drop of a few millivolts per hour can amount to scores of millivolts per day, or to hundreds of millivolts per longer time periods. Break-off of the measurements after the passage of a brief interval can lead to results that are much too high, or are crudely approximate at the very best.

We have carried out numerous experiments for the purpose of comparing the processes involved in the establishment of a potential in extended E_h measurements on moistened pulverized rocks, working first with careful protection from the air and then in open vessels. Rapid measurements in open vessels‡ again led to low negative values for reduced rocks. The results obtained here were higher by some 50 mv than those obtained in measurements carried out in the various systems described above; they were also less stable and the reproducibility was not as good. This work underscored the significance of the time required for the measurements, negative values being rarely attained in short term measurements. These results indicate that correct E_h can indeed be obtained by the methods described above [115-117], but only if the time of measurement is extended to 10-15 days.

Measurements of redox potentials on reduced rocks showed cases in which slow fall-off in the electrode potential continued for more than 7-15 days, and even some instances in which this fall-off was not completed after 45 to 60 days. The results compiled in Table 4 illustrate this point.

---

* It has already been pointed out that a potential is rapidly established in E_h measurements on oxidized rocks.
† The lowest mean value for the initial electrode readings cannot be taken as the E_h value for the medium, since at least a part of the electrode surface will have been cathodically polarized in the preliminary treatment and this will sharply reduce the potential at the beginning of measurement.
‡ Each vessel being completely filled with the rock and the electrodes lowered to the bottom of it.

TABLE 4. Redox Potential Measurements on Samples of Sandy Siltstones

Lower Buchak deposits		Upper Tsaritsyn deposits	
time from beginning of measurements, days	E, mean of potential readings on 4 electrodes, mv	time from beginning of measurements, days	E, mean of potential readings on 4 electrodes, mv
2	−104	5 hr	24
3	−127	2	−123
5	−169	6	−127
7	−199	12	−193
12	−216	19	−204
14	−224	28	−212
16	−243	60	−224
45	−250		

Redox potentials can be calculated in such cases from E_h measurements made in the course of the first 10-15 days. The method employed here is based on the fact that the measured potential (E) asymptotically approaches E_h as the time (t) increases. Equations for various curves with asymptotes are drawn on in the calculations. Since the curve showing the relation between E and t approximates a rectangular hyperbola (Fig. 9) beyond a certain point, the method of least squares can be applied to a series of such hyperbolas, each with the lines $E = E_h$ and $t = b$ as the asymptotes, i.e.,

$$(E - E_h)(t - b) = a, \tag{10}$$

or

$$E = \frac{a}{t - b} + E_h, \tag{11}$$

a, b, and E_h being unknown constants.

Definite potential readings E_1, E_2, ..., E_n are associated with the times t_1, t_2, ..., t_n. Three measurements would suffice for the determination of E_h if the representative points covering them fell exactly on the parabolas of Eq. (10), but such is far from being the case.

Let the exact values corresponding to Curve (10) be $E_1 + \varepsilon_1$, $E_2 + \varepsilon_2$, ..., $E_n + \varepsilon_n$, where ε_1, ε_2, ..., ε_n are error values which are expressed as functions of the other quantities by the equations:

$$\left. \begin{aligned} \varepsilon_1 &= \frac{a}{t_1 - b} + E_h - E_1 \\ \varepsilon_2 &= \frac{a}{t_2 - b} + E_h - E_2 \\ \bullet\ \bullet\ \bullet\ \bullet\ &\bullet\ \bullet\ \bullet\ \bullet\ \bullet\ \bullet\ \bullet\ \bullet\ \bullet \\ \varepsilon_n &= \frac{a}{t_n - b} + E_h - E_n \end{aligned} \right\} \tag{12}$$

The theory of probability shows that the most probable values of a, b, and E_h will minimize the sum of squared errors, F, where

$$\varepsilon_1^2 + \varepsilon_2^2 + \ldots \varepsilon_n^2 = F. \tag{13}$$

The function F will be at a minimum when the various partial derivatives with respect to the unknown quantities are simultaneously equal to zero, that is to say, when

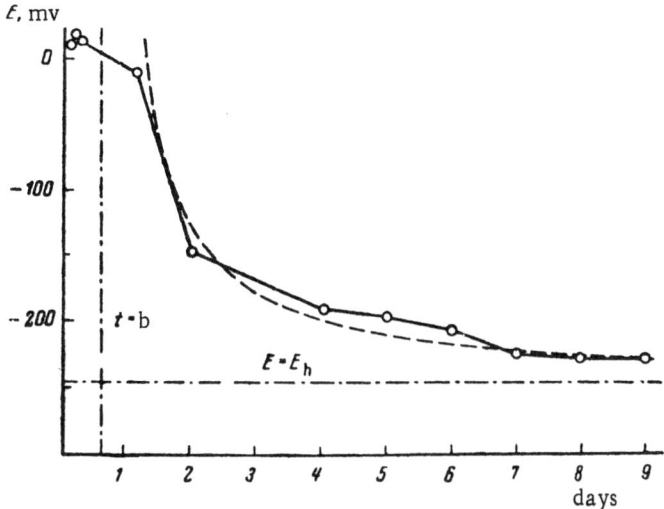

Fig. 9. Comparison of a typical curve for the establishment
of potential and an equilateral hyperbola (shown in dotted form).

$$
\left.
\begin{aligned}
\frac{\partial F}{\partial a} &= \sum_{i=1}^{n} \left(\frac{a}{t_i - b} + E_{\mathrm{h}} - E_i \right) \frac{1}{t_i - b} = 0 \\
\frac{\partial F}{\partial b} &= \sum_{i=1}^{n} \left(\frac{a}{t_i - b} + E_{\mathrm{h}} - E_i \right) \frac{a}{(t_i - b)^2} = 0 \\
\frac{\partial F}{\partial E_{\mathrm{h}}} &= \sum_{i=1}^{n} \left(\frac{a}{t_i - b} + E_{\mathrm{h}} - E_i \right) = 0 .
\end{aligned}
\right\}
\tag{14}
$$

Although an exact solution of this series of equations cannot be obtained in practice, solution does become possible if a definite value is assigned to b. The system of (14) then assumes the form

$$
\left.
\begin{aligned}
\sum_{i=1}^{n} \left(\frac{a}{t_i - b} + E_{\mathrm{h}} - E_i \right) \frac{1}{t_i - b} &= 0 \\
\sum_{i=1}^{n} \left(\frac{a}{t_i - b} + E_{\mathrm{h}} - E_i \right) &= 0 .
\end{aligned}
\right\}
\tag{15}
$$

Test shows the E_{h} values obtained by this method to be in good agreement with those actually measured. Redox potential measurements on two rock samples (Table 5) gave values of −242 and −134 mv, for example while the values calculated from points in the interval from 4 to 15 days (b = 2*) were −256 and −147 mv, respectively.

The weak point in this method of calculation lies in the fact that the work is protracted when the number of measurements is large. Results closely approximating the calculated can, however, be obtained by the following geometrical procedure. Let the E_{h}, t data be plotted in such way that the values laid off on the axis of abscissas are those of $1/(t - b)$ rather than of t itself; the curve obtained here will no longer be a hyperbola, but a straight line, and, moreover, one cutting the axis of ordinates at $E = E_{\mathrm{h}}$. The points in this new type of plot will fall on a straight line to the same degree that the points in the E_{h} vs. t plot fall on the hyperbola. The

---

*The value of b is approximated from the form of the E vs. t curve.

TABLE 5. Redox Potential Measurements on Samples of Sandy Siltstones

Time from beginning of measurements, days	Mean of potential readings on 4 electrodes, mv		Time from beginning of measurements, days	Mean of potential readings on 4 electrodes, mv	
	lower Buchak deposits	Tsaritsyn deposits		lower Buchak deposits	Tsaritsyn deposits
2	36	28	9	−234	−122
4	−170	− 61	11	−236	−128
5	−202	− 70	12	−236	−131
6	−219	− 92	14	−242	−134
7	−230	−105	15	−238	−133
8	−234	−118			

problem of solving the system of Eqs. (15) is thus reduced to the more simple problem of passing a straight line through a series of points. A construction of this kind for the measurements of Table 5 is shown in Fig. 10.

The fact that the one asymptote is selected rather arbitrarily (b = 2, for example) may cause the first few points to markedly depart from the hyperbola, thus affecting the calculated value of E_h. It would be difficult to detect a situation of this kind if the system of Eqs. (15) were to be solved algebraically, but is shows up quite clearly in the geometrical method. Thus the dotted straight line would be the best line for the upper set of points in Fig. 10, even though it does not pass through the point for the first four days.

Graphical estimation of redox potential from the data of Table 4 indicates the values finally obtained to be in good agreement with the equilibrium values when the measurements are extended over 45-60 days. Calculated E_h values which are close to the final measurements can be obtained from the results of the first 10-15 days, just as in the case of the data of Table 5.

Increase in the number of points improves the accuracy of calculations of this kind.

It is therefore recommended that measurements be made for 10-15 days when determining redox potentials of rocks. If the potential has not reached a steady value at the end of this time, the latter can be approximated from an E vs. 1/(t − b) plot in the manner shown above. Shortening of the period of measurement reduces the accuracy of determination, and is therefore not permissible at the present stage of development. The effectiveness of redox potentials on rocks can be increased by increasing the number of systems employed and carrying out simultaneous measurements on a number of samples.

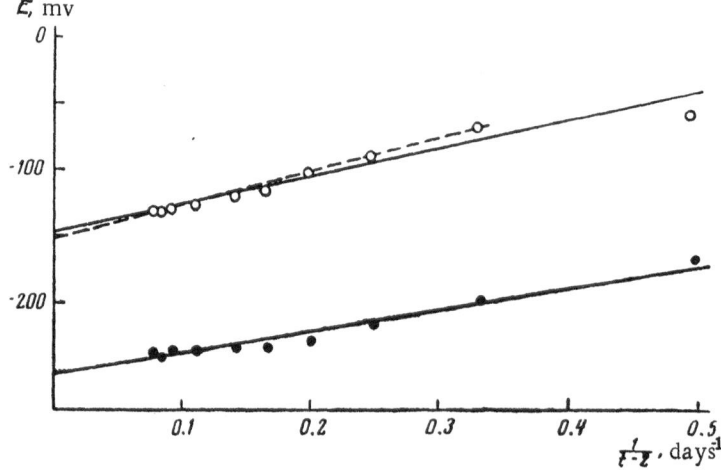

Fig. 10. Graphical calculation of E_h from the data of Table 5.

CHAPTER 6

THE SIGNIFICANCE OF THE PRELIMINARY
TREATMENT OF THE PLATINUM ELECTRODES

The rate of establishing electrochemical equilibrium in redox potential measurements depends on the redox systems present in the solution, their capacity, and their ability to impose a potential on the electrodes. The detailed behavior of the electrode is of minor significance when the capacity of the reversible redox systems of the medium is high. The importance of the characteristic redox capacity of the individual platinum electrode increases as the capacity of the solution system diminishes [108, 138]. Electrode potentials are slowly established when the redox systems of the medium have low capacity, and the potentials read from the various electrodes may then differ by scores, or even hundreds, of millivolts [107].

It was noted long ago that the surface of a platinum electrode is altered by the establishment of a potential, the resulting changes persisting even after the electrode is removed from the medium. In other words, the potential initially recorded by an electrode, and the rate at which this potential changes, both depend on the history of the electrode, more particularly, on the medium in which it has been stored and the time of storage [139, 140].

The fact that an extended period of time is required for establishing an electrode potential, and that various electrode readings differ markedly, even then, has led most workers to the conclusion that the electrodes must be given some sort of preliminary treatment. A variety of methods have been suggested for the treatment and storage of electrodes.

The treatment proposed by Bradfield, Batjer, and Oskamp [141] is quite complicated. These workers washed the electrodes with a chromate mixture, rinsed them with water and alcohol, heated them over the bunsen flame, and then once more washed them with water. This same type of treatment has been adopted by Peech and Batjer [139], by I. P. Serodobol'skii [135], and by M. M. Konova [142]. V. G. Savich [113] stored the electrodes in a concentrated NaOH solution after treatment in a chromate mixture, and washed them with water prior to use. S. V. Bruevich [143] stored the electrodes in a 20% NaOH solution, and rinsed them with solution of HCl and sea water prior to use. P. A. Kryukov [137] recommended that the electrodes be treated with a chromate mixture and then repeatedly washed with distilled water; he further proposed that the electrodes be cleaned with the soil in which field measurements were to be carried out. B. A. Neunylov [15] heated the electrodes to redness in the bunsen flame and then treated them with a chromate solution. I. G. Vazhenin [109] washed the electrodes with a hot chromate solution for 3-4 h and then washed them with water. K. A. Ovsyannikova [17] held the electrodes in a chromate solution for at least 2 h prior to measurement, washed them with distilled water, and then rubbed them dry with filter paper. E. S. Itkina [115] cleaned the electrodes by immersing them for 2 h in a chromate mixture, washed them with water, and then stored them in doubly distilled water. L. V. Pustovalov and E. I. Sokolova [116] treated the electrodes with a chromate mixture and 10% HCl, and then washed them with alcohol; working under field conditions, they rubbed the electrodes with the rock in which the measurements were to be made. Hewitt [11] cleaned the electrodes after use by immersing them in hot HNO_3. O. G. Karandeeva [144] stored the electrodes in a 12% alkali solution, washed them alternately with a 40% alkali and a 40% H_2SO_4 solution prior to use, and then rinsed them with distilled water. I. P. Merzlyakov and A. G. Silin [145] gave the electrodes no special treatment but introduced several into a soil similar to that on which measurements were to be made and then selected for use those showing only insignificantly different potential readings.

No justification has been given for any one of these methods and it is quite likely that the effect of each is simply to free the electrode of accidental contaminants such as grease films. L. Michaelis [6] has pointed to the poisoning of the electrode by previously adsorbed substances as the factor responsible for the slow approach to a potential in a medium with low-capacity redox systems. Those proposing the various treatments have advanced no explanation of this poisoning and given no proof that the treatment in question would reduce the time for establishing the potential and the divergence of the potential readings on different electrodes.

Detailed studies on the relation between the surface adsorption, on the one hand, and the capacity and potential of the platinum electrode, on the other, have been presented in various papers on electrochemistry. The results of this work are, however, almost never drawn on in measuring redox potentials of natural products.

A. N. Frumkin and his coworkers have developed a technique of studying metal surfaces, and have extensively applied it, particularly to platinum [146-151]. Here the electrode is charged by a weak current in a reaction-free solution and the potential of the electrode with respect to the solution measured as the charging takes place. So called charging curves are then constructed from the data. A typical charging curve for cathodic polarization in 1 N H_2SO_4 following preliminary polarization, or oxidation, in air is shown in Fig. 11.

A. N. Frumkin and A. I. Shlygin [146] have shown the potential drop at the platinum−solution interface to be additively composed of two components, one due to the ionic double layer and the other to the adsorbed hydrogen and oxygen. Charging curves can be used to develop the relation between the electrode potential and the surface adsorption of these last two elements. Three sections are clearly marked out on the charging curve presented above. The segment a is associated with a gradual alteration of potential and a slow removal of adsorbed atomic oxygen:

$$O_{ads} + 2H^{\cdot} + 2e' \rightarrow H_2O,$$

$$O_{ads} + H_2O + 2e' \rightarrow 2OH'.$$

The capacity of the platinum is minimal along segment b; here the electrode is free of adsorbed gases and the potential is fixed by a double layer with positive charges on the metallic surface and anions in solution. The sign of the platinum surface charge alters as the potential falls still further, the double layer now being turned with the negative side toward the metal and the cations in solution. The $H^{\cdot}$ ions begin to discharge as the potential diminishes and the platinum becomes covered with a layer of atomic hydrogen. Here the capacity of the electrode begins to rise once more (Section c).

The form of the charging curve depends on the solution composition. The potential ranges for hydrogen and oxygen adsorption overlap in alkaline solutions, for example, and hydrogen deposition begins before oxygen removal is completed [5]. B. V. Ershler and M. A. Proskurnin have studied [152] the relation between the electrode capacity and the potential.

N. I. Nekrasov has pointed out [153, 154] that each value of the electrode potential is associated with a definite degree of either hydrogen or oxygen saturation of the electrode surface in the given medium. The electrode will, for example, be covered with a layer of oxygen in strong oxidizing solutions, i.e., in solutions of high redox potential. On the other hand high potentials are always recorded when an oxygen saturated electrode is introduced into solution, regardless of whether saturation has been by anodic polarization or by contact with an oxidizing agent, or the air. A. D. Obrucheva and A. N. Frumkin give the following general description of the mechanism of establishing a potential at a platinum electrode immersed in a solution containing redox systems. "The potential of the platinum is determined by the amount of chemically bound oxygen which the electrode surface has picked up during contact with the air. Reduction of this oxygen displaces the potential in the cathodic direction, the magnitude of the displacement depending only on the electrode oxygen to reducing agent ratio."

Study of charging curves has shown [5, 156] that the amount of oxygen taken up by the platinum increases with the time of contact of the electrode with the air, eventually reaching a value several times greater than that required for the formation of a monolayer of the gas. From this it follows that the process in question is

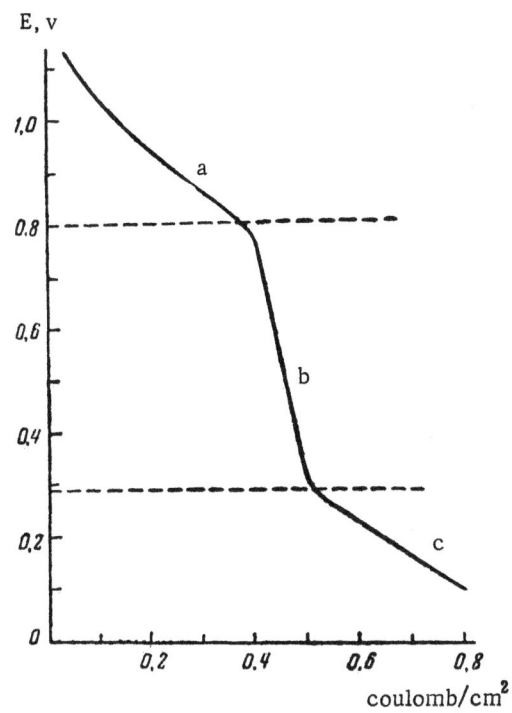

Fig. 11. Charging curve for a platinum electrode
in 1 N H_2SO_4 (taken from the paper of A. N. Frumkin,
V. S. Bagetskii, Z. A. Iofa, and B. N. Kabanov [5]).

not confined to the surface, but extends into the body of the metal as well. T. V. Kalish and R. Kh. Burshtein [157] have studied the effect of adsorbed oxygen on the electronic work function of platinum by a thermoelectric method and concluded from the results obtained that oxygen penetrates into the metal. Their calculations indicate the total oxygen adsorption by calcined platinum to be some 300 times greater than that corresponding to an atomic monolayer.

The adsorption of atmospheric oxygen by platinum is a slow process. Study of charging curves again shows that the Pt—O bond becomes stronger with the passage of time, the bond strength approaching that characteristic of oxide phases [5, 150, 151]. The oxygen which penetrates into the metal does not directly participate in the establishment of the electrode potential but is reliberated as the surface oxygen is removed. Thus, A. D. Obrucheva has pointed out [156] that oxygen liberation from the interior of the metal will cause a gradual increase in the potential of an electrode which has been subjected to cathodic polarization after long contact with the air.

Much work has been devoted to the slow oxide film formation on platinum surface in solutions of such strong oxidizing agents as $Ce^{\cdots}$, $MnO_4^{\prime}$, $Cr_2O_7^{\prime\prime}$, and HNO_3 [158-160].

The adsorbed oxygen separates only gradually from the platinum electrode to oxidize the adjacent layers of solution and is therefore one of the principal factors responsible for slow establishment of equilibrium between the electrode and the redox systems of the medium. Preliminary treatment must then not only free the electrode surface of accidental contaminants but also reduce the amount of oxygen taken up by the metal, thereby diminishing the electrode capacity.

N. I. Nekrasov [153, 154], R. Chagovets [7], and M. S. Zakhar'evskii have, among others, directed attention to the relation between the electrode potential and such surface adsorption of hydrogen and oxygen as is met in redox potential determinations. R. Chagovets [7] has noted that the electrode polarization can be insignificant at the beginning of measurements when the amount of oxygen on the electrode is high, and large when the oxygen has been removed and the potential established, especially in measurements without amplification. M. S. Zakhar'evskii has concluded on the basis of a wealth of experimental material that occluded oxygen can be just as significant as adsorbed oxygen in fixing potentials. He has introduced the concept of electrode inertia [9, 107, 108], defined as the ratio of working surface area to total volume of metal. The inertia of the electrode diminishes as this ratio increases and the electrode comes more rapidly into equilibrium with the medium.

Other data also indicate a diminution of the electrode capacity during establishment of a potential. T. I. Nekhotenova [99] has noted that the potential establishment is slow only in the first determination of the redox potential of well waters. The potential will be established rapidly in subsequent E_h measurements at other depths, if the apparatus is not brought into contact with the air between readings. I. P. Serdobol'skii [135] has made similar observations in the course of redox potential measurements on soils. We ourselves have repeatedly observed that preliminary electrode treatment is unnecessary in a series of redox potential measurements on similar rock samples where the electrodes are not allowed to remain in contact with the air between readings, potentials being established rapidly under these conditions.

Gradual evolution of previously occluded oxygen accounts for the fact that a maximum is observed at the beginning of the potential curve if the electrodes have been previously held for long periods of time in contact with the air, or treated with some sort of oxidizing agent. The fact that oxygen is evolved by the electrode more rapidly than the solution in contact with the electrode is reduced, accounts for the increase in potential over the first few hours of measurement. R. Chagovets [7] has noted that the oxygen adsorbed by platinum is a much stronger oxidizing agent than ordinary atmospheric oxygen. This conclusion is confirmed by the electrochemical work referred to above, since this has shown the surface adsorbed oxygen to be in the atomic condition.

When analyzed from this point of view, the widely employed chromate mixture treatments are seen to fail in their principal purpose of reducing the electrode capacity. Washing the electrodes with a chromate mixture is useful in eliminating organic contaminants, but it also permits further adsorption of oxygen by the platinum, especially if contact is prolonged and the mixture heated.

These conclusions are confirmed by redox potential measurements. Thus I. Martinez de Marguia [161] has shown that final washing of platinum electrodes with concentrated HNO_3 and water, or with hot chromic acid and water, will lead to high potentials which are unstable and tend to diminish with the passage of time. Peech and Batjer [139] noted that complex treatment with chromate mixture washing usually gives high potential readings, so that electrodes which have been handled in this way must be held in a soil suspension prior to measurement. I. A. Geller [138] has shown that treatment of the electrodes with a chromate mixture will increase the time required for establishing the potential by a factor of 4-5. This time can, on the other hand, be markedly reduced by treating the electrodes with a solution whose redox potential is approximately the same as that of the test soil.

The alkali treatment of electrodes proposed by certain investigators [113, 143, 144] has been cleared up by the work of A. D. Obrucheva [156]. Her study of charging curves has shown that the alkali diminishes the amount of oxygen taken up by the platinum. An analogous diminution of the strength of oxygen–metal bonding under the action of alkali has also been observed by G. A. Deborin and B. V. Ershler for gold electrodes [162]. Our own redox potential measurements on waters and sedimentary rocks [163] has shown the surface state of platinum electrodes to be considerably improved by storage in ~1-2 N alkali, the divergence between the readings of the various electrodes and the time required for the measurements both being reduced.

Other means are also available for the removal of adsorbed oxygen from platinum. Thus R. Chagovets[7] held the electrodes in Fe·· solutions for oxygen removal, and Kolthoff and Tanaka [158] used solutions of Fe·· and As··· for this same purpose. M. S. Zakhar'evskii [9] saturated the electrodes with hydrogen, holding them in overnight contact with metallic zinc in a dilute H_2SO_4 solution. This same procedure was followed by I. L. Rabotnova [25].

Cathodic polarization is one of the most effective methods of removing oxygen from electrodes. A. N. Frumkin has pointed out [148] that no other method will so effectively reduce the anodic polarization and the accompanying retention of oxygen, and, at the same time leave the metallic surface intact.

The most acceptable value of the redox potential is generally considered to be that obtained on a number of identical electrodes which have been given the same preliminary treatment. N. I. Nekrasov and O.I. Parfenova [164] have expressed the conviction that this convention does not necessarily lead to correct results. It is their opinion that a value is correct only if it is approached by several electrodes which have been deliberately chosen to have different initial states. Thus the electrodes used in their redox potential measurements on bacterial cultures were subjected to short term polarizations of opposite sign. Electrode depolarization occurs in the course of the measurements and the curves for cathodic and anodic depolarization gradually approach one another. This method gives trustworthy results, without reducing the time required for the measurements. The authors themselves point out that this method will not generally increase the consistency of the various electrode readings. Cases are often observed in which the cathodic and anodic polarization curves differ by more than 50 mv, or even cross over. Opposite signed polarization has also been used by R. Chagovets [7] and by I. A. Geller[138].

While N. I. Nekrasov and O. I. Parfenova consider that cathodic polarization has no advantages over anodic, M. S. Zakhar'evskii [9] has shown that the oxygen, and hydrogen, saturation of the electrode is not the same in these two cases, especially if reducing conditions prevail in the medium. The inertia of anodically polarized electrodes is quite high. Various electrochemical studies have also shown that cathodic and anodic polarizations are not equally useful here [5, 146, 150], oxygen saturation of the electrode under anodic polarization being much more irreversible than hydrogen saturation under cathodic polarization. This is due to the fact that the Pt–O bond is stronger than the Pt–H bond.

The results of our own work with oppositely signed electrode polarizations in redox potential measurements on sedimentary rocks are in complete agreement with the conclusions of M. S. Zakhar'evskii. This point is illustrated by Fig. 13 of Chapter 7 which shows that the potential of a cathodically polarized electrode will reach the E_h value of the medium several days earlier than an anodically polarized electrode. Cathodic polarization at potentials corresponding to the liberation of gaseous hydrogen from the electrode does not, however, reduce the time required for the measurements, which remains the same as for the untreated electrode. It has been shown in [163] that an electrode subjected to unlimited cathodic polarization will approach the E_h value of the medium just as slowly as an untreated electrode, but from the side of negative values. The potential vs. time curve for a cathodically polarized electrode can pass through a minimum as a result of gradual elimination of hydrogen from the metal with reduction of the surrounding solution.

Comparison of potential curves for cathodically polarized electrodes and for untreated electrodes suggests that the best preliminary treatment would be cathodic polarization at a potential low enough to assure removal of adsorbed oxygen but high enough to prevent the evolution of gaseous hydrogen.

Limited polarization of opposite sign was applied by I. A. Geller [16, 110, 138, 165]. He proposed that polarization be carried out in a galvanic cell consisting of one electrode in the medium of unknown redox potential and a similar electrode in a medium of either higher or lower potential. Shorting such a cell through a high resistance will cause only slight mutual separation of the electrode potentials and these potentials can therefore rapidly approach a common value without exerting any profound effect on the E_h value of the medium.

We have developed a method for limited cathodic polarization with potential control and have tested the same in measurements on sedimentary rocks. The polarization circuit is shown in Fig. 12. Prior to beginning the measurements, the potentials of the electrodes * are reduced to what would seem to be a characteristic value for the rock in question. A potentiometer is used to control the potential. Polarization is stopped after 20-30 min and the electrodes carried out of the reduction zone by being lowered 3-4 cm from the upper portion of the pulverized and moistened rock where they had been held. The potential of the anodically polarized electrode is not considered in the E_h measurements since it differs markedly from the potentials recorded by the other electrodes.

There is still no general agreement on the best choice of potential, time of polarization, current strength, and so on, for the various rocks. It can be stated, however, that controlled cathodic polarization is a very promising method for fixing the exact state of the electrode.

---

* Three electrodes connected in series were generally polarized simultaneously.

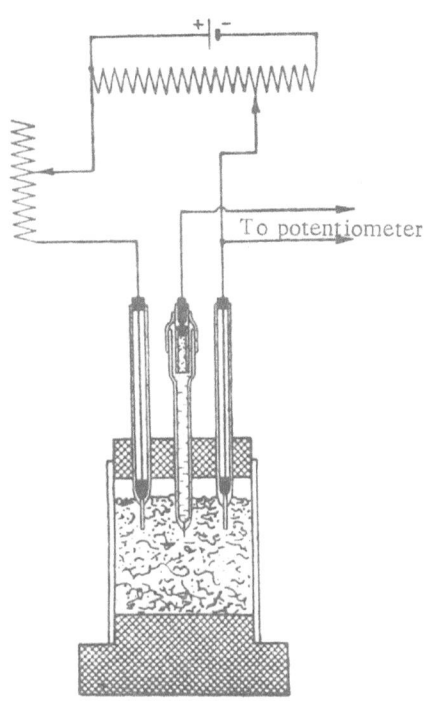

To potentiometer

Fig. 12. Circuit for controlled cathodic electrode polarization for redox potential measurements on sedimentary rocks.

It should be pointed out that the time required for redox potential measurements on reduced sedimentary rocks will be diminished by no more than 2-5 days by even the best of preliminary electrode treatments, the establishment of equilibrium between the solution and the rock minerals being the principal factor responsible for the delay in reaching a final value. It is frequently observed that the potential will rise (equilibrium being established between the electrodes and the solution) and then again fall to low values if polarization has been led to very low potentials (~-250 mv) at the beginning of measurements. To appreciably shorten the time required for measurements, it will be necessary to develop methods whereby measurements can be carried out under the natural moisture conditions. It is clear that limited cathodic polarization of the electrodes and agitation of the rock during measurement will both play a central role here.

The use of platinized glass electrodes offers one method of reducing the time required for establishing equilibrium between the electrodes and the solution. The method of preparing such electrodes has been known for a long time [166, p. 140] but the materials required for platinum plating on glass were not generally available. M. S. Zakhar'evskii [127] considerably simplified the platinizing solution with a view to obtaining low inertial electrodes. Comparison showed that the platinized electrodes had certain undoubted superiorities. Electrodes of this kind have been used by V. G. Savich [113], M. S. Zakhar'evskii [107, 108], V.A. Rabinovich, and O. V. Kurovskaya [128], and others, for redox potential measurements on various materials.

Our own measurements of redox potentials of rocks have also shown that platinized glass electrodes have very definite advantages over electrodes of the usual type [163]. Equilibrium with the rock is established much more quickly with these electrodes than with electrodes of the common type.

The low inertia of platinized glass electrodes traces back to the fact that the platinum layer is quite thin and the amount of oxygen which can penetrate into the metal correspondingly small. The mechanical weakness of the platinum layer is a drawback to their use. Even M. S. Zakhar'evskii [127] has noted that such electrodes cannot be washed with chromate solutions; experience has shown us that it is impossible to store them in either water or alkali, the platinum layer breaking down as the result of leaching of the glass. For this reason, the measurement of redox potentials in rock was continued for no more than a few days, the lifetime of the platinized glass electrodes being usually less than 1-2 measurements. Leaching proceeds more slowly in porcelain than in glass and for this reason platinized porcelain electrodes have been prepared and tested [167]. Experiment has shown that the platinum layer on porcelain is sturdier than that on glass but it too will disintegrate in the course of extended redox potential measurements. Thus the short lifetime of platinized glass or porcelain electrodes makes it difficult to apply these electrodes to redox potential measurements in rocks.

THE ACCURACY AND REPRODUCIBILITY
OF REDOX POTENTIAL MEASUREMENTS
ON SEDIMENTARY ROCKS

It has been pointed out above that questions concerning the accuracy of redox potential determinations do not usually arise in measurements on highly oxidized rocks. The situation is much more involved in the case of E_h determinations on reduced rocks where isolation from the air is required if stable results are to be obtained. Potential drop extending over several days then carries the electrode potentials to a steady E_h value which proves to be entirely reproducible in parallel measurements on the same rock sample.

Table 6 presents an example of a redox potential determination in which the measurements were continued well beyond the establishment of electrode potentials. The figures given here are representative of the stability of the lower potential values which are eventually reached and, at the same time, indicative of good sealing of the apparatus in which the measurements were carried out. Table 7 shows the reproducibility of the results of simultaneous redox potential measurements on a single specimen of rock in three different systems.

Taking the redox potential of the rock to be the lowest of the mean values of three electrode readings, the results of the E_h measurements of the Table would be −160, −172, and −154 mv. The maximum variation of these results in 18 mv, while the deviation from the arithmetical mean of the E_h values found in the three systems does not exceed 10 mv.

The maximum variation of the E_h values obtained in a series of these measurements was 41 mv (see Table 2). Such reproducibility must be considered completely satisfactory since the variation may actually trace back to differences in the oxidation states of various regions in the rock sample. V. A. Rabinovich and O. V. Kurovskaya [128] assume a reproducibility of 15 −20 mv in their measurements on soils, while K. ZoBell reported a value of 50 mv for his work on sea deposits [23].

TABLE 6. Electrode Potential Stability for Considerably Extended Measurements

Sandy siltstone from lower Buchak deposits		Sandy siltstone from upper Tsaritsyn deposits		Sandy siltstone from lower Buchak deposits	
time, from beginning of measure-ments, days	E, mean of 3 electrode readings, mv	time, from beginning of measure-ments, days	E, mean of 3 electrode readings, mv	time, from beginning of measure-ments, days	E, mean of 3 electrode readings, mv
1	68	1	35	1	98
2	36	6	−179	3	52
4	−170	12	−214	5	−53
7	−230	19	−221	6	−74
8	−234	28	−219	20	−86
11	−236	60	−224	32	−60
14	−242				
15	−238				

TABLE 7. Establishment of Potential for Simultaneous E_h Measurements on a Single Rock Sample from the Lower Buchak Deposits

Time, from beginning of measurements	Mean of potential readings on 3 electrodes, mv		
	1st experiment	2nd experiment	3rd experiment
2 hr	105	46	44
5 hr	86	36	38
20 hr	60	6	− 10
2 days	−132	−129	−141
3 days	−142	−157	−151
5 days	−160	−172	−150
7 days	−154	−161	−145
10 days	−145	−158	−154

One method for checking the validity of redox potential measurements is to use electrodes of oppositely signed polarizations. Figure 13 shows the results obtained in two experiments based on this principle. Redox potential measurements on two rock samples were continued for several days, the mean potential readings for three electrodes being −115 and −146 mv (dotted lines). One electrode in each system was then cathodically polarized, one anodically polarized, and the third left untreated.* The depolarization curves show the potentials of the polarized electrodes converging to the original values with the passage of time. The mean potential readings 25 days after polarization were −115 and −135 mv.

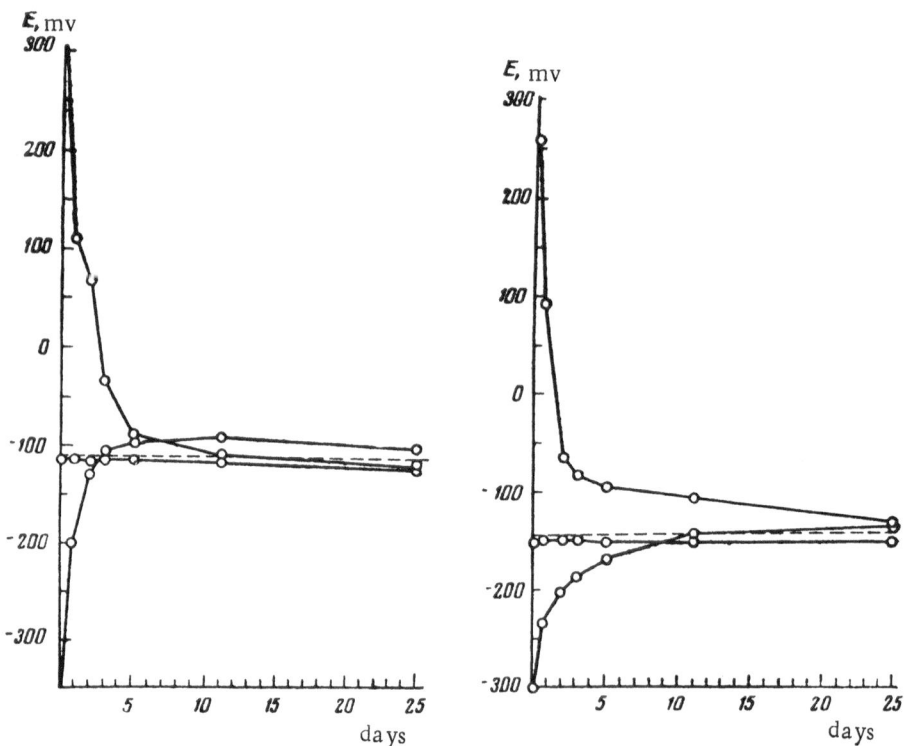

Fig. 13. Electrode depolarization curves for redox potential measurements on rock samples from the lower Buchak deposits (the dotted lines represent the mean electrode potential determined prior to polarization).

---

*These measurements were carried out in systems of type No. 2. The electrodes were polarized in the upper part of the apparatus, and then lowered by 2-3 cm.

TABLE 8. Results of Experiments on the Effect of Pyrites on Redox Potentials
of Oxidized Rocks

Time, from beginning of measurements, days	E, mean value of 2 electrode readings, mv		Time, from beginning of measurements, days	E, mean value of 2 electrode readings, mv	
	1st experiment	2nd experiment		1st experiment	2nd experiment
1	529	626	14	485	587
2	525	621	24	475	570
5	507	609	1 months	473	565
7	497	605	2 months	455	534
10	491	599	9 months	425	473

The rocks in question here contained pyrite, siderite, and glauconite. Our determined values of E_h and pH for the unoxidized samples [119] were close to those predicted for this mineralogical composition by the diagram of W. C. Krumbein and R. M. Garrels [58].

These facts indicate that redox potentials can be determined with reasonable accuracy in both reduced rocks and highly oxidized rocks. While the results obtained with reduced rocks may be too high, we consider that the error will probably be no more than 20-25 mv.

It is clear that precise values of the redox potential cannot be obtained unless the sample is carefully isolated during storage. We have used a 1 cm layer of a 1 : 1 paraffin–asphalt mixture as an isolating covering. The sample is to be isolated as soon as possible after collection.

Special attention must be given redox potential measurements on partially oxidized rocks, that is to say, on rocks containing compounds such as pyrite and ferric oxide which are characteristic oxidation and reduction forms. Measurements of this kind also lead to definite E_h values, but the results cannot be considered as even approximations to the equilibrium potentials. Sections of different degrees of oxidation are apparent to the naked eye, and the reproducibility of the results is low. Extended measurements show a gradual reduction in potential which continues for many months. This reduction can be explained as the result of slow reaction between pyrites and dissolved ferric oxide with accumulation of ferrous oxide in the mineral solutions, e.g.,

$$2Fe^{\cdots} + FeS_2 \rightarrow 3Fe^{\cdot\cdot} + 2S^0.$$

This hypothesis is supported by the experiments of Table 8, where pulverized pyrite was added to two samples of highly oxidized rock. Stable redox potentials were reached after nine months.

Further improvement in the method of determining redox potentials of sedimentary rocks must be pursued in several directions. The following three lines of attack would be the most important:

1. More complete isolation of the sample from the air, from the moment of collection to the end of measurements. The development of new types of apparatus in which all operations, including the selection of a piece of rock from the isolated sample, would be carried out under nitrogen.

2. Development of a preliminary treatment and conditions for storage which would precisely fix the initial state of the platinum electrodes. The preparation of low inertia, sturdy electrodes.

3. Carrying out the E_h determinations at the natural moisture content, making use of an apparatus in which the pulverized rock could be agitated during measurement.

_____

*There has been no detailed study of the mechanism of the pyrite oxidation by ferric oxide ions or atmospheric oxygen.

CHAPTER 8

METHODS FOR pH DETERMINATIONS
IN SEDIMENTARY ROCKS

The pH is one of the most important characteristics of a sedimentary rock. The concept of pH is similar to that of redox potential insofar as it applies to the liquid phase of rock impregnating solutions. The expression "rock pH" will be understood as designating the pH of a solution which has come into equilibrium with the rock minerals.

Ordinary glass electrodes externally reinforced by a layer of metallic alloy [168], or prepared in spear form [125], can be used in carrying out pH determinations on pulverized and slightly moistened rocks. Moistening is required to assure the necessary degree of contact between electrode and rock, but the addition of water breaks down the acid–alkali rock equilibria, just as in redox potential determinations. Passage of a definite period of time is then required in order to reestablish these equilibria.

Our own pH determinations were carried out as part of a study of the oxidation states of ground waters and sedimentary rocks [132], the first measurements being made in closed systems under an atmosphere of nitrogen at the same time as the redox potential was determined. The pH values obtained in this way were not stable, but usually fell off by 0.1-0.3 units toward the end of measurements extending over several days. There was no assurance that this reduction was solely due to renewed establishment of the acid-alkali equilibria with the rock minerals.

A series of experiments showed the pH of a rock sample held in contact with the air to be stable over a period of several hours. pH determinations on rock pastes could be carried out in open vessels since equilibrium between the solution and the glass electrode was established in several minutes. Here it was necessary to know the pH displacement resulting from moistening.

Our rock pH measurements repeatedly showed an increase resulting from the addition of water, both in closed systems and in open vessels. A. V. Trofimov [169] has also observed that the pH of a soil suspension is increased by the addition of water, this effect becoming more pronounced as the salt concentration of the soil solutions is reduced. K. ZoBell [23] has pointed to an increase in the pH of sea sludges with increasing dilution. This is clearly related to the well-known Wigner suspension effect (cited in [170]), which involves a difference in pH values between a colloidal suspension and its ultrafiltrate. B. P. Nikol'skii [171] has offered an explanation of this effect based on the Donnan theory, pointing out that the nonuniformities in ionic distribution associated with composition differences in suspension and ultrafiltrate must give rise to charged colloidal particles of limited mobility ($H^{\cdot}$ ions bound to negatively charged colloidal particles). He has also noted the existence of an inverse relation between the total concentration of ions in solution and the magnitude of the suspension effect.

This theory explains the increase in pH arising from addition of distilled water to the pulverized rock as the result of increased nonuniformity of hydrogen ion distribution in the mineral solutions.

Studies on the pH values of various mineral pastes led F. Marshak [172] to conclude that the alteration of pH in dilution is the result of ion hydrolysis during slow establishment of new solution–mineral equilibria.

Coleman, Williams, Nielson, and Jenny [173] have observed a suspension effect in ion exchange resins and various soils, and concluded that it is only apparent, actually tracing back to the use of methods which are inadequate for pH determinations in systems containing colloidal particles. These workers ascribe the effect to the existence of a potential difference in a liquid compound present in deposit and sediment, this compound being formed at the interface between the solution and the salt bridge of the calomel electrode. They have proposed replacement of the term "pH" by "apparent pH" in all studies on systems containing colloidal particles. These conclusions have not found general support, and have been subjected to criticism from various other workers (K. Marshall [174], E. Eriksson [175], etc.).

The observed effect of dilution of the rock impregnating solutions is compounded of contributions from the suspension effect, the breakdown of the acid—alkali equilibria originally existing between solution and rock minerals, and the effect of liquid compound potentials (diffusion potential).

An explanation of the pH displacement resulting from addition of water must await development of methods for determining rock pH under natural moisture conditions. Measurements of this kind have been carried out on ancient sea deposits by O. I. Dmitrenko and E. S. Zhupakhina [176]. These workers used an extremely elongated, thick walled, glass electrode (described as "unbreakable") which could be directly introduced into sludges and clays in the natural condition. These authors noted that a potential is established only slowly (15-20 min) at such an electrode; they also found that addition of water to the sludge caused an increase in the pH, the value for a 1 : 1 suspension differing from the value for a suspension with natural moisture content by 0.6 units.

Our pH measurements on sedimentary rocks with spear glass electrodes were carried out in such manner that the rock was pulverized in a mortar, transferred into a glass, and the calomel and glass electrodes then introduced directly into the loose mass. The rock around the electrodes was lightly tamped with a glass rod and the pH measured over 2-3 min, after which the electrodes were removed, the rock agitated, the electrodes reintroduced, and so on, until a potential had been completely established. Repeated introduction of the electrodes into the various regions of the rock gives the same result as diffusion in pH measurements on liquids and is necessitated by the fact that the mobility is low and changes in the hydrogen ion concentration of the rock in contact with the glass equilize out very slowly. Agitation brings the electrode into contact with fresh portions of the rock, thus helping to establish the potential much more rapidly than would otherwise be the case.

pH determinations on rocks in their states of natural moisture content can be carried out in 20-30 min by this method. The emf of the cell formed by the glass and calomel electrodes increases as the pH rises. The following experiments were carried out to show that this slow rise in emf reflected establishment of the potential of the glass electrode rather than breakdown of the equilibria existing in the rock.* The electrode was repeatedly introduced into the rock until its potential was established, and the pH then measured on freshly

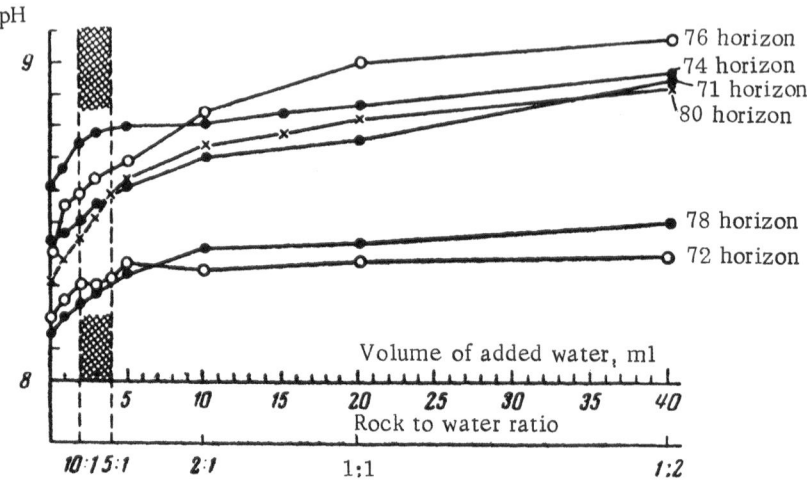

Fig. 14. Alteration of the pH resulting from the addition of water to rock samples from various stratigraphic horizons of Buchach and Tsaritsyn deposits.

---

*For example, through displacement of the carbonate equilibrium resulting from contact with the air.

prepared samples of the same rock which had been taken directly from the core and pulverized before the measurements. The results were exactly the same as those obtained in the earlier measurements, the usual variation being no more than 0.04 unit. On the other hand, rinsing the glass electrode with distilled water between the measurements markedly altered the value of the electrode potential.

The displacement of pH resulting from addition of water to various rocks was studied in a number of experiments. 20 g of the pulverized rock was placed in a small glass vessel. The potential of the glass electrode was first determined at the natural moisture content, and the rock then moistened and carefully agitated with small quantities of water, the pH being determined after each addition. The results of these experiments are shown in Fig. 14.

The data indicate that the pH of each rock involved in these studies was increased by the addition of water, the increase being particularly pronounced after introduction of the first portions of water. Addition of enough water to carry the rock to a thick paste (indicated by the hatched region in the figure) increased the pH by 0.1-0.2 units. Addition of enough water to carry the rock into a 1 : 1 rock to water suspension raised the pH by 0.6 unit.

The potential of the glass electrode is more rapidly established and proves to be more stable if the rock is moistened rather than left in its natural state, and pH measurements on rocks of the same type are best carried out on thick pastes, a correction being applied to the results. The error in pH determinations of this kind is probably less than 0.1 unit.

CONCLUSIONS

The preceding has made it clear that the direct electrometric determination of redox potentials of sedimentary rocks is entirely feasible and will lead to stable and reproducible results when precautions are taken to isolate the rock from the air and the measurements themselves are extended over a sufficiently long period of time. The methods presently available for such determinations require preliminary moistening of the pulverized rock if correct results are to be obtained. The amount of added water should be held to a minimum, however, since moistening leads to a breakdown of the solution—rock mineral equilibria.

The fact that considerable time is required for establishing a potential in E_h measurements on rocks traces back to the low capacity of the solution redox systems and to alteration in the system states in the course of the preliminary preparation. Modern methods yield high values for reduced rocks if the measurements are not continued beyond 10 days. The final E_h value can be estimated from measurements made in the course of 10-15 days, if a still longer time is required for establishing a potential.

Oxygen adsorption by the platinum electrode is the principal impediment to rapid establishment of an electrode—solution potential in redox potential measurements. The amount of adsorbed oxygen can be reduced, and the redox capacity of the electrode diminished, by treatment with alkalis and various reducing agents, and by cathodic polarization. Controlled polarization at a definite potential for a limited period of time is the most promising method for E_h determination in sedimentary rocks.

The surface state of platinum is not improved by the chromate solution treatment which has been recommended by various workers. When necessary, the electrodes should be cleaned with the chromate solution long before they are to be used in measurements, and then given further treatment to reduce the quantity of adsorbed oxygen.

An important further step in the development of methods for the determination of redox potentials of sedimentary rocks would be the discovery of procedures permitting such determinations to be carried out at the natural moisture content of the rock.

pH determinations on rocks can usually be carried out without the necessity of isolating the rock from contact with the air. Moistening the rock with distilled water leads to an increase in the pH value. The introduction of corrections makes it possible to carry out pH determinations with the rock in its natural moisture state.

LITERATURE

1. W. Latimer, The Oxidation States of the Elements and Their Potentials in Aqueous Solution [Russian translation], Moscow, IL (1954).

2. P. A. Kryukov and G. P. Avseevich, Hydrolysis and redox potentials in the ferrous-ferric system, Tr. LOVIUA 17:1935-1950 (1933).

3. W. C. Bray and A. V. Hershey, The hydrolysis of ferric ion. The standard potential of the ferric-ferrous electrode at 25°, J. Am. Chem. Soc. 56(9):1889-1893 (1934).

4. W. C. Schumb, M. S. Sherrill, and S. B. Sweetser, The measurement of the molal ferric-ferrous electrode potentials, J. Am. Chem. Soc. 59(11):2360-2365 (1937).

5. A. N. Frumkin, V. S. Bagotskii, Z. A. Iofa, and B. N. Kabanov, The Kinetics of the Electrode Processes, Izd. MGU (1952).

6. L. Michaelis, The Physiological Significance of Redox Potentials [Russian translation], Goskhimizdat (1932).

7. R. Chagovets, An electrometric study of the functionally related oxidation-reduction processes in mouse tissue, Fiziol. Zh. SSSR 22(3-4):534-544 (1937).

8. N. I. Nekrasov, The physical chemical principles of the determination of redox potentials and the significance of these potentials for biology, Usp. Eksperim. Biol. 6(3-4):3-38 (1927).

9. M. S. Zakhar'evskii, A rapid method for the determination of redox potentials in biological systems, Mikrobiologiya 9(9-10):872-878 (1940).

10. R. G. Bates, Electrometric pH Determinations. Theory and Practice, John Wiley & Sons, Inc., New York (1954).

11. L. F. Hewitt, Oxidation Reduction Potentials in Bacteriology and Biochemistry, Williams and Wilkins Co. (1950).

12. N. P. Remezov, Oxidation and reduction processes in Podzal soils, Byul. Pochvoveda No. 4-6:7-18 (1929).

13. A. D. Chernenkov, The air regime of soils and the yield of sugar beets, Khimizatsiya Sotsialist. Zemledeliya, No. 7-8:146-159 (1936).

14. I. L. Rabotnova, The redox potential of the medium and its relation to the fixation of molecular nitrogen, Mikrobiologiya 10(5):526-534 (1941).

15. B. A. Neunylov, Oxidation-reduction processes in rice field soils and methods of regulating them to increase production, in collection: Scientific Papers of the Agricultural Experiment Station of the Primorskii Region, No. 1 (1948), pp. 65-112.

16. I. A. Geller and E. G. Khariton, The effect of nitrobacteria on the redox potentials of plant tissues, Dokl. Akad. Nauk SSSR 78(5):1041-1043 (1951).

17. K. A. Ovsyannikova, The redox potentials and pH of saline waters and sludges from Lake Sakskii, Gidrokhim. Materialy 19:43-54 (1951).

18. N. N. Yurganov, Geochemical studies of sedimentary rocks for environmental analysis, Geol. Sbornik, No. 2, Tr. VNIGRI, No. 95:521-529 (1956).

19. V. E. Levenson, Problems of mud volcanism and geochemical bituminization, in collection: The Results of Studies of Mud Volcanoes of the Krymsko-Kavkazskoi Geological Province, Moscow, Izd. AN SSSR (1939), pp. 143-165.

20. V. A. Ekzertsev, Determination of the capacities of the microbiologically active layers of the sludge deposits in certain lakes, Mikrobiologiya 17(6):476-483 (1948).

21. S. F. Fedorov, Prospecting methods for oil fields and data on the formation of the latter, Tr. Inst. Nefti, Akad. Nauk SSSR 1(2):3-25 (1950).

22. N. J. Volk, The determination of redox potentials of soils, J. Am. Soc. Agr. 31(4):344-351 (1939).

23. C. E. ZoBell, Studies on redox potential of marine sediments, Bull. Am. Assoc. Petrol. Geologists 30(4):477-513 (1946).

24. W. H. Pearsall and C. H. Mortimer, Oxidation-reduction potentials in waterlogged soils, natural waters, and muds, J. Ecol. 27(2):483-501 (1939).

25. I. L. Rabotnova, The Role of Physical Chemical Conditions (pH and rH_2) in the Activity of Microorganisms, Moscow, Izd. Akad Nauk SSSR (1957).

26. M. S. Zakhar'evskii, Methods for determining the redox potentials of bacterial cultures, Zh. Mikrobiol. Epidermiol. i Immunobiol., No. 2-3:87-90 (1939).

27. M. S. Zakhar'evskii, Change in the oxidation-reduction potential in cultures of Staphylococcus aureus during penicillin bacteriostasis. Zhurnal Mikrobiol. Epidemiol. i Immunobiol., No. 11:29-33 (1948).

28. S. V. Bruevich, The chemistry of the Okhotsk Sea deposits, Tr. Inst. Okeanol. 17:41-132 (1956).

29. S. V. Bruevich and E. D. Zaitseva, The chemistry of the Bering Sea deposits, Tr. Inst. Okeanol. 26:8-108 (1958).

30. V. M. Goldschmidt, The history of the iron family minerals in nature, in collection: The Fundamental Ideas of Geochemistry, Leningrad, Goskhimizdat (1933), pp. 221-249.

31. A. E. Fersman, Geochemistry, Vol. 2, Leningrad, Khimteoret (1934).

32. L. V. Pustovalov, The Petrography of Sedimentary Rocks, Moscow-Leningrad, Gostoptekhizdat (1940).

33. A. G. Betekhtin, Hydrothermal solutions, their nature and formation processes, in collection: Basic Problems in the Study of Ore Deposits, Moscow, Izd. Akad. Nauk SSSR (1953), pp. 122-175.

34. A. P. Vinogradov, Geochemical Reactions and the Dispersion of the Elements in Soils, Moscow, Izd. Akad. Nauk SSSR (1957).

35. B. Mason, Oxidation and reduction in geochemistry, in collection: Problems of Physical Chemistry in Mineralogy and Petrography, Moscow, IL (1950), pp. 133-147.

36. H. L. James, Sedimentary facies of iron formation, Econ. Geol. 49(3):235-293 (1954).

37. K. B. Krauskopf, Separation of manganese from iron in sedimentary processes, Geochim. et Cosmochim. Acta 12(1-2):61-84 (1957).

38. R. M. Garrels, Mineral forms as functions of pH and redox potentials especially in zones of oxidation and secondary enrichment of sulphide beds, in collection: The Thermodynamics of Geochemical Processes, Moscow, IL (1960), pp. 245-266.

39. V. E. McKelvey, D. L. Everhart, and R. M. Garrels, A review of hypotheses concerning the genesis of uranium beds, in collection: The Geology of Atomic Raw Materials, Moscow, Gosgeoltekhizdat (1956), pp. 25-52.

40. V. V. Shcherbina, Redox potentials and their application to the study of mineral paragenesis, Dokl. Akad. Nauk SSSR 22(8):508-511 (1939).

41. C. A. Chapman and G. K. Schweitzer, Electrode potentials and free energy changes in geology, J. Geol. 55(1):43-47 (1947).

42. N. M. Strakhov, The diagenesis of deposits and its significance for sedimentary ore formation, Izv. Akad. Nauk SSSR, Ser. Geol., No. 5:12-49 (1953).

43. N. M. Strakhov, The diagenesis of contemporary sea and lake sludges and the conversion of the latter to sedimentary rocks, in collection: The Formation of Deposits in Contemporary Water Sources, Moscow, Izd. Akad. Nauk SSSR (1954), pp. 553-608.

44. N. M. Strakhov and E. S. Zalmanzon, The distribution of authigenic mineralogical forms of iron in sedimentary rocks and its significance for lithology, Izv. Akad. Nauk SSSR, Ser. Geol. No. 1:34-51 (1955).

45. N. M. Strakhov, Concerning the concept of diagenesis, in collection: Problems in the Mineralogy of Sedimentary Formations, Vol. 3-4, Izd. L'vovskogo Univ. (1956), pp. 7-26.

46. N. M. Strakhov, Stages in the formation of sedimentary rocks and problems involved in their study, in collection: Methods of Study of Sedimentary Rocks, Vol. 1, Moscow, Gosgeoltekhizdat (1957), pp. 7-28.

47. L. A. Gulyaeva, Geochemistry of the Devonian Terrigenous Rocks of the Ural-Volga Region, Author's abstract of her dissertation , Moscow, Institut nefi AN SSSR (1953).

48. L. A. Gulyaeva, Deposits of hydrogen sulfide basins of the geological past, Dokl. Akad. Nauk SSSR 92(5):1019-1022 (1953).

49. L. A. Gulyaeva, Geochemical indicators of oxidation-reduction conditions during sedimentation of marine terrigenous deposits, Dokl. Akad. Nauk, SSSR, 98(6):1001-1004 (1954).

50. G. I. Teodorovich, Sedimentary geochemical facies according to the profile of oxidation-reduction potential and probable oil-producing types of these facies, Dokl. Akad. Nauk, SSSR, 96(3):609-612(1954).

51. G. I. Teodorovich, Authigenous Minerals of Sedimentary Rocks, Moscow, Izd. Akad. Nauk SSSR (1958).

52. G. I. Teodorovich, The Study of Sedimentary Rocks, Leningrad, Gostoptekhizdat (1958).

53. A. S. Zaporozhtseva, The joint occurrence of glauconite and chamosite in rocks, Dokl. Akad. Nauk, SSSR, 97(5):903-906 (1954).

54. G. I. Teodorovich, Siderite geochemical facies of seas and saline waters in general as petroleum sources, Dokl. Akad. Nauk SSSR 69(2):227-230 (1949).

55. G. I. Teodorovich, Concerning oil bearing rocks, Neft. Khoz., No. 8:52-55 (1958).

56. Z. L. Maimin, Results from a study of the conditions for petroluem formation, in collection: The Conditions for Petroleum Formation, Tr. VNIGRI, No. 82:237-263 (1955).

57. A. V. Dokukin and L. S. Dokukina, Acidic Iron Waters, Their Formation and the Battle with Them, Moscow-Leningrad, Ugletekhizdat (1950).

58. W. S. Krumbein and R. M. Garrels, The origin and classification of chemical deposits, and their dependence on pH and redox potentials, in collection: The Thermodynamics of Geochemical Processes [Russian translation], Moscow, IL (1960), pp. 73-121.

59. E. V. Podol'skaya and K. F. Rodionova, The forms of sulfur and iron in pre-Devonian deposits of the central portion of the Russian platform, in collection: Problems of the Geology and Geochemistry of the Petroleum Gas Regions of the Russian Platform and the Northern Caucasus, Tr. Vses. Nauchn.-Issled. Inst., No. 4:101-116 (1954).

60. N. S. Spiro, I. S. Gramberg, and Ts. L. Vovk, The use of manganese for reconstructing the redox potential during the period of deposit formation, Tr. Nauchn.-Issled. Inst. Geol. Arktiki 98(1):90-100 (1959).

61. N. M. Strakhov, A chemical study of sedimentary rocks for genetic and correlation purposes, in collection: Methods for the Study of Sedimentary Rocks, Vol. 2, Moscow-Gosgeoltekhizdat (1957), pp. 157-186.

62. V. V. Shcherbina, The concentration and dispersion of the chemical elements in the earth's crust resulting from oxidation and reduction processes, Dokl. Akad. Nauk SSSR 67(3):507-511 (1949).

63. I. I. Romm, A geochemical characteristic of contemporary deposits of the Taman Peninsula, in collection: Contemporary Analogues of Oil Bearing Facies, Moscow-Leningrad, Gostoptekhizdat (1950), pp. 180-203.

64. I. I. Romm, Certain geochemical characteristics of sea precipitates, in collection: The Accumulation and Transformation of Organic Matter in Contemporary Sea Precipitates, Moscow, Gostoptekhizdat (1956), pp. 144-160.

65. A. B. Ronov, Organic carbon in sedimentary rocks (in connection with their petroleum content), Geokhimiya, No. 5:409-423 (1958).

66. T. I. Kazmina, The geochemical conditions for the formation of Devonian and earlier deposits of the Volga-Ural region, in collection: Conditions for Petroleum Formation, Tr. VNIGRI, No.82:68-111 (1955).

67. T. I. Kazmina, Z. L. Maimin, and Yu. N. Petrova, The problem of the conditions of sedimentation in the Devonian basin of the southwestern part of the Russian platform in terms of certain geochemical indices, No. 2, Tr. VNIGRI, No. 95, 497-510 (1956).

68. K. I. Lomot', The conditions for rock formation in the Volgo-Ural region, Tr. VNIGRI, No. 82, 24-67 (1955).

69. Z. L. Maimin, The problem of the origin of petroleum, Geol. Sbornik, No. 1, Tr. VNIGRI, No. 83:99-118 (1955).

70. K. F. Rodionova and E. V. Podol'skaya, The distribution of various forms of sulfur and iron in Devonian rocks of the central part of the Russian platform as an index of the geochemical conditions for deposit accumulation, in collection: Material on the Geology of the Regions of the Russian platform of the northern Caucasus, Tr. Vses. Nauchn.-Issled. Inst., No. 9:139-164 (1956).

71. K. F. Rodionova, E. V. Podol'skaya, and A. I. Volodchenkova, The geochemistry of the Devonian deposits of southeastern Tatary, in collection: Material on the Geology of the Region of the Russian platform of the Northern Caucasus, Tr. Vses. Nauchn.-Issled. Inst., No. 9:164-204 (1956).

72. N. N. Yurganov, Geochemical studies of sedimentary rocks of the Katangliiskii oil regions of northern Sakhalin, Geol. Sbornik, No. 2, Tr. VNIGRI, No. 95:511-520 (1956).

73. N. N. Yurganov, The problem of the establishment of petroleum producing geochemical facies from the chemical composition of rocks, in collection: Geology and Petroleum Capacity of Sakhalin, Tr. VNIGRI, No. 99:180-200 (1956).

74. N. N. Yurganov, Experimental comparison of deposits of the same age from geochemical data, Geol. Sbornik, No. 4, Tr. VNIGRI, No. 105:251-260 (1957).

75. N. K. Huber and R. M. Garrels, The relation between the formation of sedimentary iron minerals, pH, and oxidation potential, in collection: The Thermodynamics of Geochemical Processes [Russian translation], Moscow, IL (1960), pp. 161-185.

76. N. P. Remezov, The dynamics of redox potentials in podzol soils, Tr. Inst. Udobrenii, No. 77 (1930).

77. L. G. Willis, Oxidation-reduction potentials and the hydrogen-ion concentration of a soil, J. Agr. Res. 45(9):571-575 (1932).

78. M. M. Kononova, The use of redox potentials to characterize soil conditions under various types of fertilization, Pochvovedenie No. 3:365-376 (1932).

79. W. Burrows and T. C. Cordon, The influence of the decomposition of organic matter on the oxidation-reduction potential of soils, Soil Sci. 42(1):1-10 (1936).

80. N. J. Volk, The oxidation-reduction potentials of Alabama as affected by soil type, soil moisture, cultivation, and vegetation, J. Am. Soc. Agr. 31(7):577 (1939).

81. I. I. Gantimurov, The principal properties of Moscow filtration fields in relation to redox conditions, Probl. Sovetskogo Pochvovedeniya, Sb. 10:59-79 (1940).

82. G. K. Davydov, Redox conditions of podzol soils and their relation to liming, Pochvovedenie, No. 10: 615-624 (1946).

83. I. P. Serdobol'skii and P. I. Shavrygin, Redox conditions in soils of the Fergana plain, in collection: Papers on the Chemistry and Agricultural Chemistry of Soils, Tr. Poch. Inst. Akad. Nauk SSSR 31:82-90 (1950).

84. Z. A. Prokhorova, The dynamics of the nutrient regime and redox processes in soils from the Moscow River bed, Pochvovedenie, No. 1:52-61 (1957).

85. N. L. Blagovidov, V. A. Rabinovich, and I. Ya. Sell'-Bekman, The variation of the redox potential along certain soil profiles in the Leningrad oblast, Pochvovedenie, No. 6:81-85 (1957).

86. S. V. Bruevich, Hydrochemical aspects of the southern Caspian, Izv. Gos. Geograf. Obshchestva 68(1): 5-34 (1936).

87. S. V. Bruevich, Redox potentials and pH of deposits of the Barents and Kara Seas, Dokl. Akad. Nauk SSSR 19(8):635-638 (1938).

88. A. V. Trofimov, Oxidation activity and pH of the basic deposits of the Barents Sea, Dokl. Akad. Nauk SSSR 23(9):921-924 (1939).

89. V. G. Savich, A physical chemical characterization of natural water sources and deposits of the Taman peninsula, in collection: Modern Correlatives of Oil Bearing Facies, Moscow-Leningrad, Gostoptekhizdat (1950), pp. 136-180.

90. K. O. Emery and S. C. Rittenberg, Early diagenesis of California basin sediments in relation to origin of oil, Bull. Am. Assoc. Petrol. Geologists 36(5):735-806 (1952).

91. V. G. Savich, The principle aspects of the oxidation state of modern sea deposits, in collection: The Accumulation and Transformation of Organic Matter in Modern Sea Deposits, Moscow, Gostoptekhizdat (1956), pp. 92-143.

92. O. K. Bordovskii, The chemistry of the deposits of the central part of the Pacific Ocean, Tr. Inst. Okeanol. 42:107-116 (1960).

93. E. A. Romankevich and N. V. Petrov, Redox potentials and pH of the northeastern Pacific Ocean, Tr. Inst. Okeanol. 45:72-85 (1961).

94. O. V. Shishkina, Redox potentials of the upper ten-meter layer of the Quaternary Black Sea deposits, Dokl. Akad. Nauk SSSR 139(5):1218-1220 (1961).

95. K. A. Ovsyannikova, Reaction of free hydrogen sulphide with iron oxide in medicinal muds, in collection: A Study of the Curative Resources of the USSR, Moscow, Medgiz (1955), pp. 211-224.

96. K. A. Ovsyannikova and N. P. Kryuchkova, The physical chemical, and microbiological processes which develop in Sakskii mud during regeneration, in collection: Problems in the Study of the Curative Resources of the USSR, Moscow, Medgiz (1955), pp. 248-260.

97. V. V. Veber, S. G. Sarkisyan, and A. N. Gorskaya, Characteristics of the organic matter in Lake Sevan deposits, Geol. Nefti i Gaza 4(1):49-52 (1960).

98. V. M. Gortikov, E. V. Rengarten, and A. A. Goryunov, Physical chemical investigations of the Martsial'nyi springs of Karelia. From the collection: Physical Chemistry of Mineral Waters and Medicinal Muds, Moscow Biomedgiz (1937), pp. 55-86.

99. T. I. Nekhotenova, The electrometric determination of redox potentials in wells, Mikrobiologiya 7(2):186-197 (1938).

100. R. J. Allgeier, B. C. Hafford, and C. Juday, Oxidation-reduction potentials and pH of lake waters and lake sediments, Trans. Wisconsin Acad. Sci. 33:115-133 (1941).

101. P. A. Kryukov and V. M. Levchenko, Hydrogen-ion concentration and redox potential of waters, Gidrokhim. Materialy 13:237-245 (1947).

102. V. M. Levchenko, Oxidation-reduction processes in the Matsestinskii waters, Gidrokhimich. Materialy, 13:229-236 (1947).

103. P. A. Kryukov, The oxidation states of the Caucasian group of mineral waters, Gidorkhim. Materialy 14:161-182 (1948).

104. B. A. Skopintsev, Redox potentials of Black Sea waters, Dokl. Akad. Nauk SSSR 108(6):1120-1123 (1956).

105. J. G. Weart and G. E. Margrave, Oxidation-reduction potentials measurements applied to iron removal, J. Am. Water Works Assoc. 49(9):1223-1233 (1957).

106. A. I. Germanov, G. A. Volkov, A. K. Lisitsyn, and V. S. Serebrennikov, Experimental study of the redox potentials of underground waters, Geokhimiya, No. 3:259-265 (1959).

107. M. S. Zakhar'evskii, The method of determining redox potentials of bacterial cultures, Zh. Mikrobiol. Epidemiol.i Immunobiol., No. 4-5:78-85 (1944).

108. M. S. Zakhar'evskii, The determination of redox potentials of biological systems, in collection: Biological Antiseptics, Leningrad, Medgiz (1950), pp. 65-69.

109. I. G. Vazhenin, The determination of redox potentials in plants, Tr. Pochv. Inst. Akad. Nauk SSSR 31:91-95 (1950).

110. I. A. Geller, Methods of Measuring Redox Potentials in Beet Sugar Tissues (Papers on the economics, agricultural engineering, mechanization, selection, and protection of sugar beets and other cultures) 32:55-57 (1950).

111. I. A. Geller, The effect of oxygen and other oxidizing agents on the redox potentials of sugar beet tissues, Dokl. Akad. Nauk SSSR 81(2):251-254 (1951).

112. M. S. Zakhar'evskii, Physical chemical methods for the determination of the initial stages of the putrefaction of meat, Vopr. Pitaniya 8(1):35-42 (1939).

113. V. G. Savich, Measurement of Redox Potentials as a Method of Automatic Control in Air Tanks, Moscow-Leningrad, Izd. Narkomkhoza RSFSR (1940).

114. L. A. Gulyaeva and E. S. Itkina, The problem of the determination of redox potentials of caustobioliths, Dokl. Akad. Nauk SSSR 81(1):71-74 (1951).

115. E. S. Itkina, A method of determining redox potentials in rocks, Tr. Inst. Nefti Akad. Nauk SSSR 2:84-91 (1952).

116. L. V. Pustovalov and E. I. Sokolova, Methods of determining pH and E_h in sedimentary rocks, in collection: Methods of Studying Sedimentary Rocks, Vol. 2, Moscow Gosgeoltekhizdat (1957), pp. 116-127.

117. E. I. Sokolova, L. P. Listova, and A. Z. Vainshtein, Methods of determining physical chemical indices (pH and E_h) of minerals, ores, and rocks, Manuscript, Acad. Sci. of USSR, committee for the study of productive forces, mineral resources section, Moscow (1958).

118. T. I. Kazmina, Z. I. Gerasyuto, and Ts. A. Rogachevskaya, Report for 1953, Manuscript, VNIGRI Funds, cited by N. N. Yurganov, Ref. 73.

119. G. A. Solomin, Oxidation states of water and rocks in the construction zone of the Stalingrad hydroelectric station, Dissertation, Novocherkassk, Gidrokhimicheskii Institut (1960).

120. J. Juranek, Oxidation-reduction potentials of rocks. Method of analysis, Prace Ustavu Naftovy Uzykum. Publ. 26-30:37-58 (1956); Chem. Abstr. 52:4430d (1958).

121. J. Juranek and J. Fuxa, Oxidation-reduction potentials of rocks. II. A study in shallow bore holes up to 200 meters, Prace Vyzkun. Ustavu Cs. Naftovych dolu. Publ. 14(49-54):107-117 (1959); Chem. Abstr. 54:20703e (1960).

122. G. Bardossy and M. Bod, A new method for measuring the redox properties of sedimentary rocks, Geokhimiya, No. 3:247-250 (1960).

123. M. Bod and G. Bardossy, Oxidation-reduction potential of sedimentary rocks, Geofiz. Kozlemen. 8:53-72 (1959); Chem. Abstr. 54:4285d (1960).

124. N. P. Tsyba, Mineral solutions from the construction zone of the Stalingrad hydroelectric station, Dissertation, Novocherkassk, Gidrokhimicheskii Institut (1958).

125. P. A. Kryukov and V. G. Sochevanov, Methods of measuring pH with the glass electrode, Gidorkhim. Materialy 22:96-103 (1954).

126. P. A. Kryukov, pH measurements with the glass electrode, in collection: Modern Methods of Chemical Analysis of Natural Waters, Moscow, Izd. Akad. Nauk SSSR (1955), pp. 7-13.

127. M. S. Zakhar'evskii, Metallized glass electrodes, Zavodsk. Lab. No. 5-6:647 (1940).

128. V. A. Rabinovich and O. V. Kurovskaya, Use of platinized glass electrodes for field determinations of redox potentials of soils, Pochvovedenie, No. 4:78-80 (1953).

129. P. A. Kryukov, A new type of calomel half-cell, Zavodsk. Lab. 6(12):1495 (1937).

130. S. Ya. Vainbaum, Electrode stability in redox potential measurements, Zavodsk. Lab. 18(4):502 (1952).

131. S. Ya. Vainbaum, A probe for E_h measurements in gas well faces, in collection: Field and Industrial Geochemistry, Vol. 1, Moscow-Leningrad, Gostoptekhizdat (1953), pp. 12-14.

132. P. A. Kryukov, G. A. Solomin, V. E. Goremykin, N. P. Tsyba, V. I. Manikhin, and E. M. Lebedeva, The oxidation states of waters and rocks from the construction zone of the Stalingrad hydroelectric station, Gidrokhim. Materialy 31:142-163 (1961).

133. P. A. Kryukov and G. A. Solomin, Methods of measuring redox potentials of waters and rocks, Gidrokhim. Materialy 28:215-221 (1959).

134. G. A. Solomin, Apparatus for measuring redox potentials of sedimentary rocks, Gidrokhim. Materialy 31:209-210 (1961).

135. I. P. Serdobol'skii, The effect of moisture on oxidation-reduction processes in podzol soils, Pochvovedenie, No. 7:47-59 (1940).

136. N. P. Remezov, Physical Chemical Methods of Studying Soils, Moscow-Leningrad, Sel'kolkhozgiz (1931).

137. P. A. Kryukov, Electrochemical methods of studying soils, in collection: Modern Methods of Studying the Physical Chemical Properties of Soils, Vol. 2, Moscow-Leningrad, Izd. Akad. Nauk SSSR (1947), pp. 16-79.

138. I. A. Geller, The measurement of redox potentials of microbiological media, Mikrobiol. Zh. Akad. Nauk Ukr.RSR 11(2):79-91 (1949).

139. M. Peech and L. P. Batjer, A critical study of the methods for measuring oxidation-reduction potentials of soils with special reference to orchard soils, Cornell Univ. Agr. Exp. Sta. Bull., No. 625:1-23 (1935).

140. S. D. Elek and E. S. Boatman, Time-lag in E_h potentials, Nature 172(4388):1056 (1953).

141. R. Bradfield, L. P. Batjer, and J. Oskamp. The significance of the oxidation-reduction potentials in evaluating soils for orchard purposes, Cornell Univ. Agr. Exp. Sta. Bull., No. 592:1-27 (1934).

142. M. M. Kononova, Study of the new formation of humous substances, Pochvovedenie, No. 10:456-470 (1944).

143. S. V. Bruevich, Redox potentials and pH values of Polar Sea soils, Probl. Arktiki, No. 3:210-224 (1943).

144. O. G. Karandeeva, A new electrode vessel for the electrometric determination of redox potentials and pH of biological liquids, Laboratornoe Delo, No. 1:50-51 (1958).

145. I. P. Merzlyakov and A. G. Silin, The measurement of redox potentials of prairie peats, Pochvovedenie, No. 10:1330-1335 (1938).

146. A. N. Frumkin and A. I. Shlygin, Concerning the platinum electrode, Dokl. Akad. Nauk SSSR 2(3):173-179 (1934).

147. B. V. Ershler, G. A. Deborin, and A. N. Frumkin, Concerning the platinum electrode. The adsorption of hydrogen and oxygen on platinum at elevated temperatures, Izv. Akad. Nauk SSSR, Ser. Khim. No. 5:1065-1073 (1937).

148. A. N. Frumkin, Electrochemical methods for studying catalysts' surfaces, Zh.Fiz.Khim. 14(9-10):1200-1207 (1940).

149. B. V. Ershler, The Passivity of Platinum, Vol. 2, Moscow, Izd. Akad. Nauk SSSR (1943), pp. 52-66.

150. Ts. I. Zalkind and B. V. Ershler, Oxygen adsorption on platinum in polarization from charging curves, Zh. Fiz. Khim. 25(5):565-576 (1951).

151. V. I. Nesterova and A. N. Frumkin, Concerning the platinum electrode, IX. Oxygen adsorption on platinized platinum resulting from contact with molecular oxygen and anodic polarization, Zh. Fiz. Khim. 26(8):1178-1183 (1952).

152. B. V. Ershler and M. A. Proskurin, The capacity of the smooth platinum electrode in various electrolytes and its relation to the electrode treatment, Zh. Fiz. Khim. 8(5):689-695 (1936).

153. N. I. Nekrasov, Irreversible redox potentials, Mikrobiologiya 6(7):849-864 (1937).

154. N. I. Nekrasov, Irreversible redox potentials, Zh. Fiz. Khim. 11(1):84-98 (1938).

155. A. D. Obrucheva and A. N. Frumkin, The mechanism of establishing a potential difference between platinum and silver salt solutions, Dokl. Akad. Nauk SSSR 4(1):11-13 (1936).

156. A. D. Obrucheva, The platinum electrode. X. An electrochemical study of the adsorption of oxygen on smooth platinum, Zh. Fiz. Khim. 26(10):1448-1457 (1952).

157. T. V. Kalish and R. Kh. Burshtein, The effect of oxygen adsorbed on platinum on contact potential differences, Dokl. Akad. Nauk SSSR 81(6):1093-1096 (1951).

158. I. M. Kolthoff and N. Tanaka, Rotated and stationary platinum wire electrodes. Residual current-voltage curves and dissolution patterns in supporting electrolytes, Anal. Chem. 26(4):632-636 (1954).

159. J. W. Ross and I. Shain, Oxidation of platinum electrode in potentiometric redox titrations, Anal. Chem. 28(4):548-551 (1956).

160. J. K. Lee, R. N. Adams, and C. E. Bricker, Studies of gold, platinum, and palladium indicating electrodes in strong oxidizing aqueous solutions, Anal. Chim. Acta 17(3):321-328 (1957).

161. J. A. T. Martinez de Marguia, Study of electrode potentials, Inst. Espan. Oceanog., Notas Resumenes Ser. II., No. 140:10 (1947); Chem. Abstr. 42:6675f (1948).

162. G. A. Deborin and B. V. Ershler, The polarization capacity of the smooth gold electrode, Zh. Fiz. Khim. 14(5-6):708-716 (1940).

163. G. A. Solomin, The preliminary treatment of electrodes for redox potential measurements, Gidrokhim. Materialy 28:222-229 (1959).

164. N. I. Nekrasov and O. I. Parfenova, Electrode depolarization curves and their use in characterizing the redox properties of aqueous media, Mikrobiologiya 7(2):164-185 (1938).

165. I. A. Geller, The effect of plant cultures on the redox regime of soils, Pochvovedenie, No. 10:920-926 (1952).

166. V. Ostval'd, R. Lyuter, and K. Druker, Physical Chemical Measurements, Chapt. 1, Leningrad, Goskhimizdat (1934).

167. G. A. Solomin and S. S. Zavodnov, Platinized porcelain electrodes, Manuscript, Novocherkassk, Gidrokhimicheskii Institut (1961).

168. P. A. Kryukov and A. A. Kryukov, Metallized glass electrodes, Zavodsk. Lab. 6(5):619-621 (1937).

169. A. V. Trofimov, Soil reactions (pH) as a function of moisture and soil solution concentration, Pochvovedenie 26(2):5-45 (1931).

170. P. A. Kryukov, Phase potentials arising in suspension creep, Tr. Pochv. Inst. Akad. Nauk SSSR 15:274-286 (1947).

171. B. P. Nikol'skii, The theory of the Wigner and Pal'man suspension effects, Pochvovedenie, No. 9:138-143 (1939).

172. F. Marshak, pH studies of mineral suspensions, Kolloidn. Zh. 12(1):41-49 (1950).

173. N. T. Coleman, D. E. Williams, T. R. Nielsen, and H. Jenny, On the validity of interpretation of potentiometrically measured soil pH, Soil Sci. Soc. Am. Proc. 15:106-114 (1950).

174. C. E. Marshall, Soil Sci. Soc. Am. Proc. 15:110 (1950).

175. E. Eriksson, The significance of pH, ion activities, and membrane potentials in colloid systems, Science 113(13):418-420 (1951).

176. O. I. Dmitrenko and E. S. Zhupakhina, A method of measuring pH of soils and rocks at the natural moisture content with the unbreakable glass electrode, Pochvovedenie, No. 1:111-123 (1957).